T0240182

Lecture Notes in Electrical Engineering

Volume 712

The book series *Lecture Notes in Electrical Engineering* (LNEE) publishes the latest developments in Electrical Engineering—quickly, informally and in high quality. While original research reported in proceedings and monographs has traditionally formed the core of LNEE, we also encourage authors to submit books devoted to supporting student education and professional training in the various fields and applications areas of electrical engineering. The series cover classical and emerging topics concerning:

- Communication Engineering, Information Theory and Networks
- Electronics Engineering and Microelectronics
- Signal, Image and Speech Processing
- Wireless and Mobile Communication
- Circuits and Systems
- Energy Systems, Power Electronics and Electrical Machines
- Electro-optical Engineering
- Instrumentation Engineering
- Avionics Engineering
- Control Systems
- Internet-of-Things and Cybersecurity
- Biomedical Devices, MEMS and NEMS

For general information about this book series, comments or suggestions, please contact leontina.dicecco@springer.com.

To submit a proposal or request further information, please contact the Publishing Editor in your country:

China

Jasmine Dou, Associate Editor (jasmine.dou@springer.com)

India, Japan, Rest of Asia

Swati Meherishi, Executive Editor (Swati.Meherishi@springer.com)

Southeast Asia, Australia, New Zealand

Ramesh Nath Premnath, Editor (ramesh.premnath@springernature.com)

USA, Canada:

Michael Luby, Senior Editor (michael.luby@springer.com)

All other Countries:

Leontina Di Cecco, Senior Editor (leontina.dicecco@springer.com)

** Indexing: Indexed by Scopus. **

More information about this series at http://www.springer.com/series/7818

Hyuncheol Kim · Kuinam J. Kim
Editors

IT Convergence and Security

Proceedings of ICITCS 2020

 Springer

Editors
Hyuncheol Kim
Namseoul University
Cheonan, Korea (Republic of)

Kuinam J. Kim
Kyonggi University
Suwon-si, Korea (Republic of)

ISSN 1876-1100   ISSN 1876-1119   (electronic)
Lecture Notes in Electrical Engineering
ISBN 978-981-15-9356-7   ISBN 978-981-15-9354-3   (eBook)
https://doi.org/10.1007/978-981-15-9354-3

This Springer imprint is published by the registered company Springer Nature Singapore Pte Ltd.
The registered company address is: 152 Beach Road, #21-01/04 Gateway East, Singapore 189721,
Singapore

Organizing Committee

General Chair

Hyuncheol Kim, Namseoul University, South Korea

Steering Committee

Nikolai Joukov, New York University and modelizeIT Inc, USA
Borko Furht, Florida Atlantic University, USA
Bezalel Gavish, Southern Methodist University, USA
Kin Fun Li, University of Victoria, Canada
Kuinam J. Kim, Kyonggi University, Korea
Naruemon Wattanapongsakorn, King Mongkut's University of Technology Thonburi, Thailand
Xiaoxia Huang, University of Science and Technology Beijing, China

Publicity Chairs

Minsu Kim, Kyonggi University, Republic of Korea
Suresh Thanakodi, National Defence University of Malaysia, Malaysia

Financial Chair

Jongmin Kim, Kyonggi University, Republic of Korea

Publication Chair

Hyeunchul Kim, Namseoul University, Republic of Korea

Programme Chair

Kuinam J. Kim, Institute of Creative Advanced Technologies, Science and Engineering, Korea
Nakhoon Baek, Kyungpook National University, Republic of Korea

Organizers & Supporters

Institute of Creative Advanced Technologies, Science and Engineering (iCatse)
Korean Industry Security Forum (KISF)
Namseoul University, South Korea
Kyonggi University, Republic of Korea
Hongik University, Republic of Korea
University of Canterbury, New Zealand
Kyungpook National University, Republic of Korea
Korea Institute of Science and Technology Information (KISTI)

Programme Committee

Yanling WEI, KU Leuven, Belgium
Prof. Ch. Satyananda Reddy, Andhra University, India
Terje Jensen, Telenor
Dr. Ahmad Kamran Malik, COMSATS University Islamabad (CUI), Islamabad, Pakistan
Pavel Loskot, Swansea University, UK
Ad Lalit Prakash Saxena, Applied Research Section, ComboConsultancy, Obra Sonebhadra UP India, India
Mauro Gaggero, National Research Council of Italy, Italy
Wing Kwong, Hofstra University, USA
Nguyen Dinh Cuong, Nha Trang University, Vietnam
Iickho Song, Korea Advanced Institute of Science and Technology, South Korea
Ong Thian Song, Multimedia University Malaysia, Malaysia
Pr. Mainguenaud Michel, Institut National des Sciences Appliquées—Normandy, France
Mohd. Saifuzzaman, Daffodil International University, Bangladesh
Fu-Hau Hsu, National Central University, Taiwan
Baojun M. A., Shanghai International Studies University, China
Harikumar Rajaguru, Bannari Amman Institute of Technology, Sathyamangaalam, India
Hyoungshick Kim, Sungkyunkwan University, South Korea
Prof. Zbigniew Leonowicz, Wrocław University of Science and Technology, Wroclaw, Poland
Hyunsung Kim, Kyungil University, South Korea
Sangseo Park, The University of Melbourne, Australia
Pao-Ann Hsiung, National Chung Cheng University, Taiwan
Noraini Che Pa, Universiti Putra Malaysia (UPM), Malaysia
Sandro Leuchter, Hochschule Mannheim University, Germany
Vasily Tarasov, IBM Research
Wun-She Yap, Universiti Tunku Abdul Rahman, Malaysia

Dr. Shitala Prasad, Agency for Science, Technology and Research
Dr. Chittaranjan Pradhan, Kalinga Institute of Industrial Technology (KIIT) Deemed to be University, India
Pr. Pascal Lorenz, University of Haute Alsace, France
Naoufel Kraiem, Sultan Qaboos University, Oman
Dr. Ramadan Elaiess, University of Benghazi, Libya
Bok-Min Goi, Universiti Tunku Abdul Rahman (UTAR), Malaysia
Zeyar Aung, Khalifa University of Science and Technology, UAE
Maurantonio Caprolu, Hamad Bin Khalifa University, Qatar
Ing. Lorenzo Ricciardi Celsi, The Sapienza University of Rome, Italy
Maicon Stihler, Centro Federal de Educação Tecnológica de Minas Gerais—CEFETMG, Brazil
Daniel B.-W. Chen, Monash University, Australia
So Jeong Kim, National Security Research Institute, South Korea
Dr. Ng Hui Fuang, Department of Computer Science, Universiti Tunku Abdul Rahman, Malaysia
Shamim H. Ripon, East West University, Bangladesh
Min-Shiang Hwang, Asia University, Taiwan
Dr. Christof Ebert, Vector Consulting Services
Stelvio Cimato, Università degli studi di Milano, Italy
Vasco N. G. J. Soares, Instituto de Telecomunicações/Instituto Politécnico de Castelo Branco, Portugal
Marco Listanti, Sapienza University of Rome, Italy
William Emmanuel Yu, Ateneo de Manila, Ateneo de Manila
Oscar Mortagua Pereira, University of Aveiro, Portugal
Dr. Somlak Wannarumon Kielarova, Naresuan University, Phitsanulok, Thailand
Kittisak Jermsittiparsert, Social Research Institute, Chulalongkorn University, Thailand
Yadigar Imamverdiyev, Institute of Information Technology, Azerbaijan National Academy of Sciences, Azerbaijan
Dr. Anooj P. L., Al Musanna College of Technology, Oman
Partha S. Mallick, Vellore Institute of Technology, India
Prof. Dr. Zuriati Ahmad Zukarnain, Universiti Putra Malaysia, Malaysia
Abdulaziz Mohammed alshareef, King Abdulaziz University, Saudi Arabia
Roy Abi Zeid Daou, Lebanese Geman University, Lebanon
Dr. Srinivas Sethi, LMCSI, Indira Gandhi Institute of Technology Sarang, India
Ljiljana Trajkovic, Simon Fraser University, Canada
Dr. Ruchi Tuli, Jubail university college, Saudi Arabia
Dr. Azeddien M. Sllame, University of Tripoli, Libya
Shyamala Doraisamy, Universiti Putra Malaysia, Malaysia
Ivan Homoliak, Brno University of Technology, Czech Republic

Preface

This LNEE volume contains the papers presented at the iCatse International Conference on IT Convergence and Security (ICITCS 2020), which was held on August 19th–21st, 2020.

ICITCS2020 will provide an excellent international conference for sharing knowledge and results in IT Convergence and Security. The aim of the Conference is to provide a platform to the researchers and practitioners from both academia, as well as industry, to meet and share cutting-edge development in the field.

The primary goal of the conference is to exchange, share and distribute the latest research and theories from our international community. The conference will be held every year to make it an ideal platform for people to share views and experiences in IT Convergence and Security related fields.

On behalf of the Organizing Committee, we would like to thank Springer for publishing the proceedings of ICITCS2020. We would also like to express our gratitude to the 'Programme Committee and Reviewers' for providing extra help in the review process. The quality of a refereed volume depends mainly on the expertise and dedication of the reviewers. We are indebted to the Programme Committee members for their guidance and coordination in organizing the review process, and to the authors for contributing their research results to the conference.

Our sincere thanks to the Institute of Creative Advanced Technology, Engineering and Science for designing the conference web page and also spending countless days in preparing the final programme on time for printing. We would also like to thank our organizing committee for their hard work in sorting our manuscripts from our authors.

We look forward to seeing all of you next year's conference.

Editors of ICITCS2020
Kuinam J. Kim
Hyuncheol Kim

Cheonan, Republic of Korea
Suwon-si, South Korea

Contents

Computer Vision and Applications

Internet of Things

Security and Privacy

Other Related Topics

Part I
Machine Learning and Deep Learning

DiagnoseNET: Automatic Framework to Scale Neural Networks on Heterogeneous Systems Applied to Medical Diagnosis

John Anderson Garcia Henao, Frédéric Precioso, Pascal Staccini, and Michel Riveill

Abstract Determining an optimal generalization model with deep neural networks for a medical task is an expensive process that generally requires large amounts of data and computing power. Furthermore, the complexity of the programming expressiveness increases to scale deep learning workflows over new heterogeneous system architectures for training each model and efficiently configure the computing resources. We introduce DiagnoseNET, an automatic framework designed for scaling deep learning models over heterogeneous systems applied to medical diagnosis. DiagnoseNET is designed as a modular framework to enable the deep learning workflow management and allows the expressiveness of neural networks written in TensorFlow, while the DiagnoseNET runtime abstracts the data locality, micro batching and the distributed orchestration to scale the neural network model from a GPU workstation to multi-nodes. The main approach is composed through a set of gradient computation modes to adapt the neural network according to the memory capacity, the workers' number, the coordination method and the communication protocol (GRPC or MPI) for achieving a balance between accuracy and energy consumption. The experiments carried out allow to evaluate the computational performance in terms of accuracy, convergence time and worker scalability to determine an optimal neural architecture over a mini-cluster of Jetson TX2 nodes. These experiments were performed using two medical cases of study, the former dataset is composed by clinical descriptors collected during the first week of hospitalization of patients in the Provence-Alpes-Côe d'Azur region; the second dataset uses a short ECG records between 30 and 60 s, obtained as part of the PhysioNet 2017 Challenge.

J. A. Garcia Henao (✉) · F. Precioso · M. Riveill (✉)
Laboratoire I3S, Université Côte d'Azur, 06900 Biot, SA, France
e-mail: henao@i3s.unice.fr

M. Riveill
e-mail: michel.riveill@unice.fr

F. Precioso
e-mail: frederic.precioso@unice.fr

P. Staccini
Université Côte d'Azur, CHU Nice, 06000 Biot, NCE, France
e-mail: pascal.staccini@unice.fr

© The Author(s), under exclusive license to Springer Nature Singapore Pte Ltd. 2021
H. Kim et al. (eds.), *IT Convergence and Security*, Lecture Notes
in Electrical Engineering 712, https://doi.org/10.1007/978-981-15-9354-3_1

1   Introduction

Determine an optimal generalization model with Deep Neural Networks (DNN) in healthcare research is an expensive training process, due to the cost of hardware and electricity or the cloud compute time and its carbon footprint required to fuel HPC systems [12]. Therefore, physicians and scientists require alternative High-Performance Computing (HPC) systems to exploit the Artificial Intelligence (AI) methods applied in medical diagnosis within hospitals to accelerate healthcare research as well as to preserve the patient data privacy with an affordable cost of hardware and electricity. In this context, the motivation of this research is to develop a programming framework to improve the usability, portability and scalability of deep learning workflows over heterogeneous systems and, evaluate low-consumption computing architecture with minimal infrastructure requirements [1], to accelerate clinical-risk-predictive models [10, 14] with an efficient balance between accuracy and energy consumption.

Based on these challenges, this paper analyses the deep learning algorithmic complexity in terms of accuracy, convergence time and worker scalability for training two different neural networks (MLP and CNN) on a mini-cluster with 14 Jetson-TX2 nodes, applied to predict the medical care purpose of hospitalized patients and to atrial fibrillation classification for cardiac arrhythmia diagnosis. Which main contribution is an open-source framework called DiagnoseNET, designed into independent and interchangeable modules for scaling deep learning models over heterogeneous system architecture applied to develop medical risk prediction models. Which increases the developer's productivity, facilitating the programming process to build and fine-tune a DNN, while its run-time abstracts the data locality, the micro batching and the distributed orchestration to scale the DNN model from a GPU workstation to multi-nodes.

2   Background and State of the Art

The main challenge is to minimize the execution time, increase the worker scalability and exploit the computing power of each hybrid processor on-chip (CPU&GPU) with 8GB of host memory capacity by each Jetson TX2. Where the data locality, the communication protocol and the coordination training modes are the key factors for efficient task mapping over the resources but it increases the programming complexity for model training and computing orchestration.

The common data-distributed methods are Bosen [13] and Federated Learning [3, 9], the approaches that use iterative-convergent Machine Learning (ML) algorithms for training. They can be applied generically to any ML method if data samples are independent and identically distributed (i.i.d.). The Bosen platform provides a distributed version for a number of well-known ML algorithms (for example, Deep Learning, Sparse Coding, K-means clustering, Random forests or Multi-class Logistic Regression), while Federated Learning is designed to be efficient in setups with a large number of users and unreliable or slow connections. Final classification or prediction models represent a weight matrix that is stored across a large number of clients. Local weight matrix is calculated in the initial step and refined over the rounds, where updates are based on the exchange of parameters with local neighbors or a single master node.

Model-distributed approaches such as Strads platform [13] require ML specialized systems that perform a partition of ML algorithms into a set of parallel tasks, in general scheduled by master node(s) and executed by a set of workers. Schedulers' task is to separate the problem into a non-overlapping set of sub-problems, divide a workload and synchronize the updates amongst the workers. This setup admits non-conflicting model updates that lead to convergence. Numerous algorithms can be deployed in this framework, such as Latent Dirichlet Allocation, Matrix Factorization, Support Vector Machine or Deep Learning algorithm based on Caffe, called Poseidon, to name a few.

Model and data-distributed algorithms for classification and prediction problems. In the literature, there exist only a few works. A hybrid distributed platform known as Angel [7] appropriately combines data partitioning, scheduling and parameter synchronization tasks and demonstrates accuracy improvement in comparison with a Petuum-based data or model distribution. There exist a number of calculus-parallelization methods, such as FlexFlow [6]. It is a hybrid data and model parallel (non-distributed) approach worth of exploring in a distributed setup, because it performs automated search of parallelization strategies that incorporates data, attribute, parameter and operator parallelization for DNN algorithms.

3 Materials and Methods

3.1 DiagnoseNET

DiagnoseNET was designed to harmonize the deep learning workflow and to automatize the distributed orchestration to scale the neural network model from a GPU workstation to multi-nodes. Figure 1 shows the schematic integration of the DiagnoseNET modules with their functionalities. The first module is the deep learning model graph generator, which has two expression languages: a *Sequential Graph* API designed to automatize the hyperparameter search and a *Custom Graph* which support the TensorFlow expression codes for sophisticated neural networks. The second module is the data manager, compose by three classes designed for splitting, batching and multi-task any dataset over GPU workstations and multi-nodes computational platforms. The third module extends the enerGyPU monitor for workload characterization, constitute by a data capture in runtime to collect the convergence tracking logs and the computing factor metrics; and a dashboard for the experimental analysis results [4]. The fourth module is the runtime that enables the platform selection from GPU workstations to multi-nodes whit different execution modes, such as

Fig. 1 DiagnoseNET
framework scheme

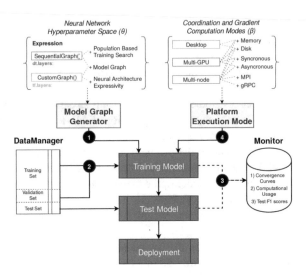

synchronous and asynchronous coordination gradient computations with gRPC or MPI communication protocols.

DiagnoseNET Model Graph Generator, In *Sequential Graph* the first step, defines the stacked layers and sets the type of each layer, their neurons numbers, the number of layers and followed by a linear output on top, since the cross entropy will be used as loss function and include the softmax function. Then the neural network hyperparameters are defined as shown in the expression 1. to generate the model graph object. *Custom Graph* uses *tf.layers* to defines the staked layers and the similar expression as the former is used to define the optimizer and loss function for generating the model graph object.

Code Example 1.1 Model definition to generate several graphic-model objects.

```
import diagnosenet as dt
stacked_layer_1 =  [dt.Relu(14637, 2048),
                    dt.Relu(2048, 1024),
                    dt.Relu(1024, 1024),
                    dt.Linear(1024, 14)]

model_1 = dt.sequentialGraph(
                input_size=14637, output_size=14,
                layers=stacked_layer_1,
                loss=dt.CrossEntropy,
                optimizer=dt.Adam(lr=0.001))
```

DiagnoseNET Data Manager, This manages the dataset according to the computational architecture, creating an isolated sandbox for each dataset and its transformations in the training process to guarantee the data location. In which, the dataset is splitting into well balance batches over the number of workers, and its worker-batch is micro batching according to the memory or parameter, as shown in the following code expressions:

Code Example 1.2 Dataset splitting and micro-batching over the workers.

```
data_config_1 = dt.Batching(
                dataset_name="medical_D1", valid_size=0.05,
                devices_number=4, batch_size=128)
```

DiagnoseNET for Distributed Training with gRPC, It harmonizes the computational resources with the dataset manager to train previously defined models over a multi-node platform, automating the gRPC communication protocol to coordinate the workers with asynchronous gradient computations. In which, the resource manager divide the dataset equally onto the workers nodes of the system where each worker has a copy of the neural network (graph) along with its local weights. Each worker operates on a unique subset of the dataset and updates its local set of weights. These local weights are shared across the cluster to compute a new global set of weights through an accumulation algorithm.

Code Example 1.3 Distribted orchestration with GRPC asynchronous.

```
import diagnosenet as dt
dt.between_graph_replication(
        d_replica_path=/myworkspace,
        d_replica_name="GRPC_replica.py",
        ip_ps="host1",
        ip_workers="host2,host3,host4,host5",
        num_ps=1, num_workers=4)
```

On the side of the replica script, is gives the model graph object, create the dataset batching, and pass both of these to a *Distributed_GRPC* object. This object is responsible for launching the experiment, through its function *asynchronous_training*.

Code Example 1.4 GRPC asynchronous replica.

```
platform = dt.Distibuted_GRPC(
        model=model_1,
        datamanager=data_config_1,
        monitor=enerGyPU(machine_type="arm"),
        max_epochs=20,
        ip_ps=argv[0], ip_workers=argv[1])

platform.asynchronous_training(
        dataset_path=/myworkspace/datasetpath,
        inputs_name="X.npy", targets_name="Y.npy",
        job_name=argv[0], task_index=argv[1])
```

DiagnoseNET for Distributed Training with MPI, DiagnoseNET implements synchronous and asynchronous MPI methods to improve performance in the communication between workers. For example, asynchronous gradient updates were optimized with a parameter called weighting, which is responsible to determine the number of workers required in each step to compute the new weights and broadcast it.

Code Example 1.5 MPI Platform Execution Modes.

```
platform = dt.Distibuted_MPI(
        model=model_1, datamanager=data_config_1,
```

```
monitor=enerGyPU(machine_type="arm"),
max_epochs=20, early_stopping=3])

platform.asynchronous_training(
    dataset_name="medical_D1",
    dataset_path=d/myworkspace/datasetpath,
    inputs_name="X.npy", targets_name="Y.npy",
    weighting=1)
```

The specifications of the MPI algorithms are described in the Appendices 1 and 2.

4   Case Studies and Neural Architectures

Medical Care Purpose Classification for Inpatients: The clinical dataset was derived from a program for medicalisation of information systems (PMSI) collection of synthetic medical information in a standardized and anonymized format from hospitalizations that carried out in activities in medical care or rehabilitation settings. In which the patient-feature composition module was used to generate the representations of the patients status in the first week of hospitalization using from one-year of the PMSI data collection. The main clinical descriptors used was demographics, admission details, hospitalization details, physical dependence, cognitive dependence, rehabilitation time, comorbidities, morbidity and etiology. The clinical dataset obtained has 116, 831 different inpatients and 14, 637 clinical-features embedded in a document-term sparse matrix [5]. In this paper we worked with the high-level group called Clinical Major Category (CMC) obtaining 14 labels-categories to classify the medical care of patients hospitalized as shown in Table 1.

Atrial Fibrillation Classification for Cardiac Diagnosis: The ECG dataset was obtained from the 2017 PhysioNet Challenge The dataset was already labeled and the four labels are: Normal, Atrial Fibrillation, Others and Noisy. The Others label means recordings of those similar heart diseases. The total number of source dataset is 8, 528. Each sample is a single short ECG lead recording. Since the length of the sample is inequivalent, samples are transformed into structured input. The position of the peaks R of recordings are extracted to get the centred windows of 260 time steps, which are complete ECG rhythms for a cycle. To better represent the behaviour of the recording, each five consecutive centred windows are concatenated into a training sample as shows in the following Table 2.

Multilayer Perceptron Network: In DiagnoseNET the network architecture was composed dynamically through fully-connected layers, each neuron is connected to all neurons of the previous layer building a stacked neural network and followed by a softmax layer on top $h_i = f(\sum_{j=1}^{n} w_{ij}x_j + b_{ij})$, where x_j is the output of the previous layer and w_{ij} is the weight value associated with x_j with a bias associated $b_{i,j}$ and n is the number of neurons in the previous layer, while f is the as activation function. Having as a baseline the neural network used in the work called *improving*

Table 1 Medical target 1: care purpose description labels

Class	Labels description	Train	Valid	Test
0	Other Situations	4,489	267	515
1	Proceedings of medical cardiovascular/respiratory care	18,299	1,122	2,263
2	Circulatory system disorders	13,074	764	1,504
3	Proceedings of neuro-muscular medical care	5,375	295	640
4	Proceedings of medical care mental health	2,929	175	344
5	Proceedings sensory and skin medical care	8,273	456	950
6	Proceedings of rheumatics/orthopedic medical care	18,080	1,061	2,106
7	Proceedings of post-traumatic medical care	14,174	801	1,619
8	Proceedings of medical amputations	741	45	89
9	Palliative care	2,056	114	256
10	Placement expectation	299	20	40
11	Rehabilitation	2,261	114	268
12	Proceedings of nutritional medical care	9,240	415	1,144
13	No grouping	16	1	3

Table 2 Medical target 2: cardiac arrhythmia labels

Class	Label description	Source dataset	Training dataset	Small samples
0	Normal	5,050	34,303	4,241
1	AF	738	6,542	815
2	Others	2,456	18,986	2,424
3	Noisy	284	1,382	171

palliative care with deep learning [2] and after finetune it to classify the medical care purpose with PACA inpatients as shown in the Appendix 4. The model used to evaluate the scalability was comprised by an input (of 10,833 dimensions), 4 hidden layers (each 512 dimensions) and a softmax output layer. As activation function was used rectified linear unit (ReLU), as loss function was used categorical cross-entropy and Adam as optimizer [8].

Convolutional Neural Network: The neural network baseline for the second medial task is based on a Convolutional Neural Network (CNN) designed to take as input the time-series of ECG signal and generates the sequence of label predictions as outputs [11]. The general neural architecture is composed using DiagnoseNET with 75 layers of convolution followed by a fully-connected layer and a softmax layer on top, as shows in the Appendix 4. The major elements in the CNN model are the residual network, the convolutional layers and the regularization methods, such as batch normalization, dropout and activation which are used to improve the performance and regularization of the CNN model. The convolutional layers are used in order to extract features relative to the form of the traces wave.

5   Experiments and Results

The experiments were conducted using the DiagnoseNET self expression codes for training the medical care purpose classification, and for training the atrial fibrillation classification. This implementations was processing over a set of gradient computation modes as synchronous or asynchronous for each communication protocol GRPC or MPI.

HPC System and Environment: For processing the distributed experiments were built a mini-cluster of 14-nodes Jetson TX2 interconnected by 1 $GigE$ switch Ethernet. The nodes are identical, independent machines and each one runs a separate OS. Every node is composed of one developer kit Jetson TX2, which contains a hybrid processor Nvidia Denver with one ARM Cortex-A57 quad-core with one a Pascal GPU 256-$CUDA@cores$ with a maximum, it has $8GB$ of LPDDR4 memory, $59.7GB/s$ of memory bandwidth and $32GB$ of internal storage.

Worker Scalability for Training the Medical Task 1: The baseline got 11.04 h as convergence time for training the MLP model described in the previous session, which was performed using gRPC asynchronous to coordinate and compute the gradient updates between 2 workers and 1 master. As shown in the Fig. 2, the best setting reduces the convergence time to 1.3 h, using MPI asynchronous to coordinate and compute the gradient updates between 12 workers and 1 master.

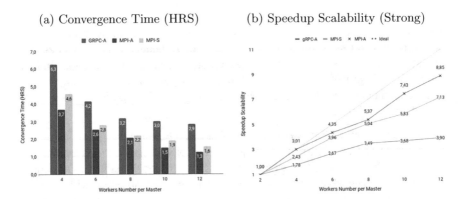

Fig. 2 Worker scalability comparison for distributed training on a mini-cluster of Jetson TX2 to classify the medical care purpose

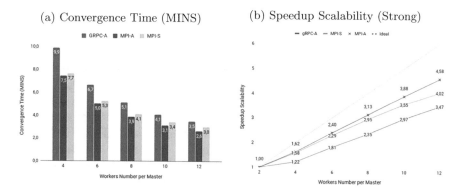

Fig. 3 Worker scalability comparison for distributed training on a mini-cluster of Jetson TX2 to the classify atrial fibrillation

Worker Scalability for Training the Medical Task 2: For the atrial fibrillation classification was used a small dataset (77MB) with 8, 528 patients and a medium model with 72 layers fully-connected of convolutional neural network with residual network connections. Where the baseline uses a gRPC asynchronous training modes with 4 workers take 13 min as a time to solution achieving one accuracy of 0.63 F1-score, while the MPI asynchronous training modes with 12 workers take 5 min as a time to solution achieving the same accuracy of 0.63 F1 score, as shown in the Fig. 3.

6 Conclusions

DiagnoseNET increases developer's productivity facilitating the programming process to build and finetune Deep Learning workflows, while its runtime abstracts the data locality and the distributed orchestration to scale each model from a GPU workstation to multi-nodes. Furthermore, implement a mini-cluster of Jetson TX2 nodes presents a good scalability for distributed training of each neural network by their medical task. Therefore, clusters with embedded computation platforms can be used as a deep learning platform system with minimal infrastructure requirements and low power consumption, offer the computing capacity for processing considerable datasets and models in the HPDA ecosystem. In the way to characterize the deep learning tasks and improve the balance between accuracy, convergence time and worker scalability, MPI asynchronous gradient computations with data parallelism offer an efficient distributed neural network training for early convergence. Likewise adapting the number of records by batch and the model dimensionality helps to minimize the bottleneck of data transfer from host memory to device memory reducing the GPU idle status.

Appendix 1: DiagnoseNET MPI Synchronous Algorithm

The algorithm 1 describes the MPI synchronous coordination training with parameter server. It uses the nodes ranks to assign them the role of parameter server or worker, defined the rank 0 as parameter server (PS) and the other ranks as workers. When launching the program, the PS does necessary pre-processing tasks, such as loading the dataset and compiling the model. After these tasks, the PS sends the model to the workers, which are ready to receive it. At each training step, the PS sends a different subset of the data to every worker to be used for loss optimization. At the end of an epoch, the PS will gather the new weights from every worker. Workers receive the collection of weights and compute the average weight for the global update. For the other computing parts, it works as the desktop version.

Algorithm 1 Synchronous MPI Kernel

---

while $ConvergenceCondition$ **do**
    if $master == True$ **then**
        for all $worker \in workers$ **do**
            $masterGrads \leftarrow received(workerGrads)$
        $averageGrads \leftarrow average(masterGrads)$
        $send(averageGrads)$
    else
        $workerGrads \leftarrow compute(model, batches)$
        $send(workerGrads)$
    if $master == True$ **then**
        for all $worker \in workers$ **do**
            $masterLoss \leftarrow received(workerLoss)$
        $averageLoss \leftarrow average(masterLoss)$
        if $overfitting(averageLoss) True$ **then**
            $send(averageLoss, earlyStopping)$
        else
            $send(averageLoss, False)$
    else
        $workerWeights \leftarrow received(masterWeights)$
        $projection \leftarrow model.Apply(workerWeights)$
        $workerLoss \leftarrow computeLoss(projection, labels)$
        $send(workerLoss)$

Appendix 2: DiagnoseNET MPI Asynchronous Algorithm

The algorithm 2 allows training multiple model replicas in parallel on different nodes with different subsets of the data. Each model replica processes a mini-batch to compute gradients and sends them to the parameter server which apply a function (mean, weighted average) between previous and received weights, then updates the global weights accordingly and send them back to the workers. In fact, every worker will compute its gradients individually until its convergence; the convergence occurs when we start having overfitting, which means that the training loss is decreasing while the validation loss increased. The master who is responsible for computing the weighted average of received weights and its own weights, will stop when all workers converge. To check the status of convergence of workers, the master has a queue that stores converged workers and when its length is equal to the number of workers, the master knows that all workers converged and stops training. Since each node computes gradients independently and does not require interaction among each other, they can work at their own pace and have greater robustness to machine failure.

Algorithm 2 Asynchronous MPI Kernel

---

while $ConvergenceCondition$ **do**
  if $master == True$ **then**
    $convergeFlag \leftarrow received(workerCond)$
    $masterGrads \leftarrow received(workerGrads)$
    $collectGrads \leftarrow collection(masterGrads)$
    $averageGrads \leftarrow average(collectGrads)$
    $send(averageGrads)$
  else
    if $overfitting(averageLoss) True$ **then**
      $send(averageLoss, earlyStopping)$
    else
      $send(averageLoss, False)$
    if $decrease(averageLoss) True$ **then**
      $send(Updated(masterWeights))$
    $workerGrads \leftarrow compute(model, workerInput)$
    $send(workerGrads)$
  if $master False$ **then**
    $workerWeights \leftarrow received(masterWeights)$
    $projection \leftarrow model.Apply(workerWeights)$
    $workerLoss \leftarrow computeLoss(projection, labels)$

Appendix 3: Hyperparameter Search to Classify the Medical Task 1

A model space contains (d) hyperparameters and (n) hyperparameters configurations defined in Table 3 and the Table 4 shows the models by number of parameters. We have established some fixed hyperparameters and decided to tune the number of units per layer, the number of layers and batch size, which are the hyperparameters that directly affect the computational cost. Each model was trained using *Adam* as an optimizer with a maximum of 40 epochs and as a loss function is used the *Cross Entropy*.

Table 3 Search space model descriptors

Hyperparameters (d)	Hyper. Configurations (n)	State
Learning rate	Adaptive L.R. starting	From: 0.001
Activation function	relu, tanh, linear	Fixed: relu
Num. Units per layer	16, 32, 64, 128, 256, 512, 1024, 2048, 4096	Search
Num. hidden layers	2, 4, 8, 16	Search
Regularization	Dropout: 0.6, 0.7, 0.8	Fixed: 0.8
Batch size	24.576, 12.288, 6.144, 3.072, 1.536, 768	Search
Num. of workers	4, 6, 8, 10, 12	Search

Table 4 Model dimension space in number of parameters (millions)

		Numbers of layers			
		2	4	8	16
Neurons by layer	16				0.24
	32			0.47	0.48
	64		0.95	0.97	1.0
	128	1.89	1.93	1.99	2.12
	256	3.82	3.95	4.21	4.74
	512	7.76	8.29	9.34	11.44
	1024	16.05	18.15	22.35	
	2048	34.2	42.6		
	4096	76.8			

According with the model dimension showed in the Table 4, we are found that is possible divided the models by Fine, middle and course grain. In which, the Fig. 4 shows that middle-grain models from 1.99 to 8.29 millions of parameters have a fast convergence in validation loss, and high accuracy levels for the majority of the 14 care purpose labels, in comparison with the other models who present a great variation in accuracy and spent more epochs to convergence.

Appendix 4: ECG Neural Architecture to Classify the Medical Task 2

The pure CNN model leads to the problem that the last layer of the model may not exploit the original features or the ones extracted in the first layers. The Fig. 5 shows the ECG neural architecture implemented using DiagnoseNET framework, which key architecture factor are the residual network connections to solve the information

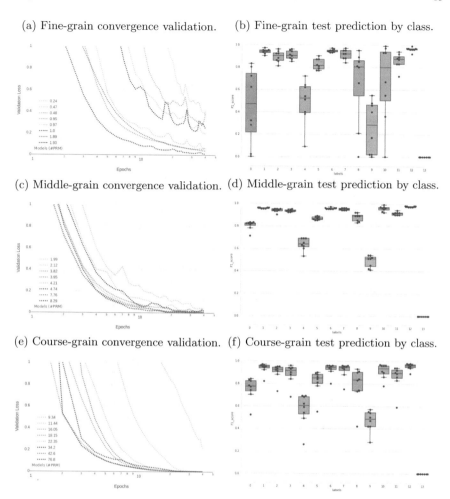

(a) Fine-grain convergence validation.   (b) Fine-grain test prediction by class.

(c) Middle-grain convergence validation. (d) Middle-grain test prediction by class.

(e) Course-grain convergence validation. (f) Course-grain test prediction by class.

Fig. 4 Experiment results for training a feed-forward neural network, using the hyperparameter model-dimension space

loss problem into the deep layers. To implement this, a second information stream is added in the model. In this way, deeper layers have access to the original features, in addition to the information processed by the previous layers. What else, two different types of residual block are included to access the different states of the information. The normal residual block preserves the size of the input while the sub-sampling residual block lowers the size of the input down to a half. By using max pooling, the network extracts only the high values from an input so that the size of its output is halved.

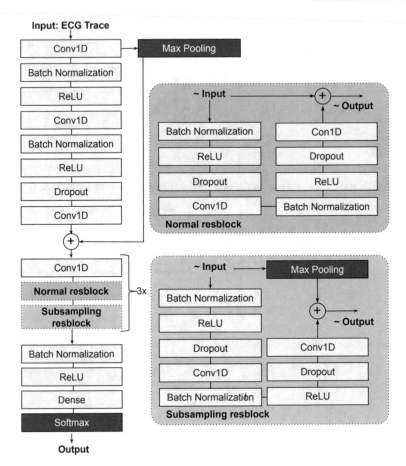

Fig. 5 ECG convolutional neural architecture

Acknowledgements We thank DU Ziqing, Mohamed Younes, Arno Gobbin and the IADB team for their help. This work is partly funded by the French government labelled PIA program under its IDEX UCAJEDI project (ANR-15-IDEX-0001). The PhD thesis of John Anderson Garci´a Henao is funded by the French government labelled PIA program under its LABEX UCN@Sophia project (ANR-11-LABX-0031-01).

References

1. Asch M, Moore T et al (2018) Big data and extreme-scale computing: pathways to convergence-toward a shaping strategy for a future software and data ecosystem for scientific inquiry. Int J High Perform Comput Appl 32:435–479
2. Avati A, Jung K, Harman S, Downing L, Ng AY, Shah NH (2017) Improving palliative care with deep learning. CoRR. arXiv:1711.06402

3. Bonawitz K, Eichner H, Grieskamp W, Huba D, Ingerman A, Ivanov V, Kiddon C, Konecný J, Mazzocchi S, McMahan HB, Overveldt TV, Petrou D, Ramage D, Roselander J (2019) Towards federated learning at scale: system design. CoRR. arXiv:1902.01046
4. Garcia Henao JA, Esteban Hernandez B, Montenegro CE, Navaux PO, Barrios Hernández CJ (2016) enerGyPU and enerGyPhi monitor for power consumption and performance evaluation on Nvidia Tesla GPU and Intel Xeon Ph
5. Garcia Henao JA, Precioso F, Staccini P, Riveill M (2018) Parallel and distributed processing for unsupervised patient phenotype representation. In: Latin America high performance computing conference. https://hal.archives-ouvertes.fr/hal-01885364, Sept 2018
6. Jia Z, Zaharia M, Aiken A (2018) Beyond data and model parallelism for deep neural networks. CoRR. arXiv:1807.05358
7. Jiang J, Yu L, Jiang J, Liu Y, Cui B (2017) Angel: a new large-scale machine learning system. Natl Sci Rev 5(2):216–236. https://doi.org/10.1093/nsr/nwx018
8. Kingma DP, Ba J (2014) Adam: a method for stochastic optimization. arXiv:1412.6980
9. Konecný J, McMahan HB, Yu FX, Richtárik P, Suresh AT, Bacon D (2016) Federated learning: strategies for improving communication efficiency. CoRR. arXiv:1610.05492
10. Maharlou H, Niakan Kalhori SR, Shahbazi S, Ravangard R (2018) Predicting length of stay in intensive care units after cardiac surgery: comparison of artificial neural networks and adaptive neuro-fuzzy system. Healthc Inform Res 24(2):109–117. https://doi.org/10.4258/hir.2018.24.2.109. http://europepmc.org/articles/PMC5944185
11. Rajpurkar P, Hannun A, Haghpanahi M, Bourn C, Ng A (2017) Cardiologist-level arrhythmia detection with convolutional neural networks
12. Strubell E, Ganesh A, McCallum A (2019) Energy and policy considerations for deep learning in NLP. arXiv:1906.02243, Jun 2019
13. Xing EP, Ho Q, Dai W, Kim JK, Wei J, Lee S, Zheng X, Xie P, Kumar A, Yu Y (2015) Petuum: a new platform for distributed machine learning on big data. IEEE Trans Big Data 1(2):49–67
14. Ye C, Wang O, Liu M, Zheng L, Xia M, Hao S, Jin B, Jin H, Zhu C, Huang CJ, Gao P, Ellrodt G, Brennan D, Stearns F, Sylvester KG, Widen E, McElhinney DB, Ling X (2019) A real-time early warning system for monitoring inpatient mortality risk: prospective study using electronic medical record data. J Med Internet Res 21(7):e13719–e13719. https://www.ncbi.nlm.nih.gov/pubmed/31278734, https://www.ncbi.nlm.nih.gov/pmc/articles/PMC6640073/

Effective Disease Prediction on Gene Family Abundance Using Feature Selection and Binning Approach

Thanh-Hai Nguyen, Tan-Tai Phan, Cong-Tinh Dao, Dang-Vinh-Phuc Ta, Thi-Ngoc-Cham Nguyen, Nguyen-Minh-Thao Phan, and Huynh-Ngoc Pham

Abstract Metagenomic is now a novel source for supporting diagnosis and prognosis human diseases. Numerous studies have pointed to crucial roles of metagenomics in personalized medicine approaches. Recent years, machine learning has been widely deploying in a vast amount of metagenomic research. Usually, gene family data are characterized by very high dimension which can be up to millions of features. However, the number of obtained samples is rather small compared to the number of attributes. Therefore, the results in validation sets often exhibit poor performance while we can get high accuracy during training phrases. Moreover, a very large number of features on each gene family dataset consumes a considerable time in processing and learning. In this study, we propose feature selection methods using Ridge Regression on datasets including gene families, then the new obtained set of features is binned by an equal width binning approach and fetched into either a Linear Regression and a One-Dimensional Convolutional Neural Network (CNN1D) to do prediction tasks. The experiments are examined on more than 1000 samples of gene family abundance datasets related to Liver Cirrhosis, Colorectal Cancer, Inflammatory Bowel Disease, Obesity and Type 2 Diabetes. The results from the proposed method combining between feature selection algorithms and binning show significant improvements in both prediction performance and execution time compared to the state-of-the-art methods.

Keywords Gene family abundance · Disease prediction · Metagenomic · Feature selection

T.-H. Nguyen (✉) · T.-T. Phan · C.-T. Dao · D.-V.-P. Ta · T.-N.-C. Nguyen · N.-M.-T. Phan · H.-N. Pham
College of Information Communication of Technology, Can Tho University, Can Tho, Vietnam
e-mail: nthai@cit.ctu.edu.vn

© The Author(s), under exclusive license to Springer Nature Singapore Pte Ltd. 2021 19
H. Kim et al. (eds.), *IT Convergence and Security*, Lecture Notes
in Electrical Engineering 712, https://doi.org/10.1007/978-981-15-9354-3_2

1   Introduction

Metagenomics (Environmental Genomics, Ecogenomics or Community Genomics) is directly the study of communities of microbial organisms in their natural environments by applying of the modern genomic techniques [1]. The application of metagenomic sequence information will facilitate the design of better culturing strategies to link genomic analysis with pure culture studies. Over the past 20 years, the development of information technology has supported metagenomics analysis, human genome research and genome analysis of pathogenic microorganisms, leading to antibiotic research in the world. At the same time, with the Next Generation Sequencing (NGS) [2] technique, the human genome has been decoded, detecting the rare Crohn disease mutations that have been identified and sought to prevent. Currently, metagenomic is a potential new data source to be applied in supporting primary care and diagnosis for human health. As described in Ehrlich, this data source can assist in diagnosing diseases, forecasting and detecting risks that can lead to people suffering from diseases, and monitoring. treatment progress [3].

A gene family is a set of genes derived by duplication of a single original gene and normally with similar biochemical functions [4]. They reflect the consequences of natural selection, random genetic propensity and molecular mechanisms of gene duplication. Gene prediction in fact still faces many challenges because many genes in a gene family are still unable to determine its function.

In this study, we propose a feature selection on datasets including gene families, then the new obtained set of features is binned by an equal width binning approach and fetched into Ridge Regression and two types of model for training: Fully Connected Neural Network (FC) and (CNN1D) to do prediction tasks. The number of features of reflect on datasets are up to more than one million features. Through six gene families abundance datasets related to various diseases in proposed investigation that shows promising prediction results.

2   Related Work

Human microbiome in health and disease plays a significant role that has recently been given considerable observation [5]. Maja and et al. demonstrate this approach combined with machine learning, to classify microbiome samples in human gut according to the pathological condition diagnosed in the human host. In addition, predicting disease-relevant features in microbial gut metagenomes by using the principle of utilizing the prokaryotic translational optimization effect combined with the machine learning based classification and enriched gene dataset that explore a supportive method to analyzing metagenomic datasets [6]. Some methods have been applied feature selection method to metagenomic datasets, for example "Subset selection based on information-theoretic" [7]. Other studies, mRMR (Min Redundancy Max Relevance) algorithms have been applied [8], Lasso [9], Elastic Net

[10], browsing and selection algorithms to reduce the number of elements of input data [8]. In addition, Hilal et al., they also use methods such as Conditional Mutual Information Maximization (CMIM), Fast Correlation Based Filter (FCBF), mRMR and eXtreme Gradient Boosting (XGBoost) [11]. A abundance of feature selection (FS) approaches, most of them is increasing as a demand to analyze data of really high dimension, normally hundreds or thousands of variables. In Gene Expression Microarray (GEM) analysis, which is also known as Differentially Expressed Genes (DEGs) discovery, gene prioritization or biomarker discovery, they investigated filter feature selection methods for informative feature detection [12]. Machine learning plays important role that help analyze Biological and Medical data. The tool was developed in combination with four classification methods (e.g. SVM, RF, Lasso and ENet) that have been widely applied in many different fields including computational biology and genome [13]. Classifiers have been implemented using the Scikit-learn (sklearn) package, the Python language [14].

3 Method

From the original gene family abundance, the data is filtered by feature selection and then is binned by an unsupervised binning method before fetching into a learning architecture such as either Linear Regression or CNN1D. The proposed method keeps implementing standard learning steps using k-fold cross-validation with some modification at such crucial points as feature selecting period and data transformation, which will promisingly produce better results after training process. After splitting considered dataset into folds, at every iteration we always have a training set and a test set. Since number of attributes in dataset are very large, we suggest a reduction method using feature selection based on Ridge algorithm learned from training set. The feature selection method will result in a finite number of features based on which training set and test set are filtered at step 4. We also suggest a binning method called EQW to apply on our data, which finds breaks on new training data at step 5 and apply the breaks at step 6. Training algorithms come into play at step 7 using two types of models: Linear Regression and Convolutional Neural Network. Evaluation at step 8 is conducted after training done.

3.1 Feature Selection on Gene Family Abundance

For the datasets with more than one million features such as gene family abundance, we proposed several approaches of Feature Selection (FS) to obtain efficient prediction tasks on different diseases. Feature selection is a really significant technique in machine learning to find the most relevant features for the predictive model from millions of genomes dataset which are strongly correlated in the data without much loss of information. Feature selection approaches how to choose a subset, variables or list of attributes that build models for describing data. Its purpose to consist of dimensionality reduction, irrelevant and redundant features removement, the amount of data required for learning [15]. So that, the selection of features plays greater importance role in the high-dimensional datasets because the number of variables is bigger than the number of monitoring.

For our considered datasets, the number of features is higher than the number of samples, which consume much memory and take a lot of training time for any learning algorithm. In order to deal with the problem, some solutions are proposed to limit the number of features so that the learning accuracy can be improved as well as memory usage and time consumption are reasonable. To achieve results in this paper, we conducted many experiments using feature selection following 3 methods: **Ridge Regression (Ridge), Lasso Regression (Lasso), Low Variance Filter (LVF)** with number of selected features equals to 1024 and two types of model for training: FC and CNN1D.

3.2 Feature Selection Combined to an Unsupervised Binning Approach

Authors in [16] attempted to using binning approaches to enhance the performance for metagenomic data. These techniques of binning purpose to reduce the effects of minor observation errors, and to get rid of noise in the data. Previous studies have applied the binning methods to bacterial species abundance. In this study, we employ the binning method on gene family abundance where the number of features is over a thousand of times compared to the number of features in species abundance.

Equal width binning (EQW) method is leveraged to discretize continuous values into discrete bin value. For the ranging of binning, we consider the minimum and maximum values in training set to generate bin values. Let denote a range for binning as [Min, Max]. In the case, we use 5 bins and Min $= 0$, Max $= 1$, hence: the width of interval is 0.1, let say w $= 0.1$. The breaks for binning are **min + w, min + 2 * w, ..., min + (k − 1) * w** which are equivalent to 0.1, 0.2, ..., 0.9, respectively.

After desired number of features are selected, their values are discretized in finite number of bins (the number of bins in this research is set to 5). The processing of generating breaks is done with learning from training set.

3.3 Learning Algorithms for the Learning

The performances of methods are measured by Average Accuracy on 10-cross-validation on test sets. The models used in experiments include Linear Regression and a shallow CNN1D. CNN1D consists of one convolutional layer (the size of 3) with 64 filters followed by a max pooling of 2 (stride 2). Linear Regression is a fully connected layer (FC). FC and CNN1D receive samples in 1D as input and produce an output which reveals the probability of having the disease by a sigmoid function. All networks have the same Adam optimization function, use a default learning rate of 0.001, a batch size of 16. In order to reduce overfitting issues, we also implement Early Stopping technique with an epoch patience of 5.

4 Dataset Benchmark

Medical studies have shown that metagenomic data collected in the human body contribute a large part to the causes of diseases. In this study, we investigate and evaluate our proposed method on six gene family abundance datasets (as shown in Table 1) that include inflammatory bowel diseases (IBD) [17], liver cirrhosis (CIR) [18], colorectal cancer (COL) [19], obesity (OBE) [20] and Type 2 Diabetes (T2D) [21] and WT2 (T2D in Europe women) [22]. These data were downloaded from curatedMetagenomicData [23] in R which were generated by HMP Unified Metabolic Analysis Network (HUMAnN2) [24]. The numbers of features of all datasets are up to more than one million of features. The total abundance of the genes is summing to Table 1.

Table 1 Information on six considered gene family abundance datasets

Dataset name	CIR	COL	IBD	OBE	T2D	WT2
#features	1,747,534	1,796,274	1,730,384	1,519,375	1,690,774	1,415,610
#samples	232	121	110	253	344	96

Table 2 Performance comparison between our proposed method (FS with Ridge and equal width binning) and the study in [25] using FC model. "ᵃ" denotes the performance in [25] using dimensionality reduction such as PCA (Principal Component Analysis) and RD_PRO (Random Projection). "ᵇ" expresses one of the best results of species abundance using gray images in [25]. The bold results are better than [25] with raw data (formatted as italic text)

Method	CIR	COL	IBD	OBE	T2D	WT2	AVG
PCAᵃ [25]	0.547	0.604	0.775	0.648	0.514	0.540	0.605
RD_PROᵃ[25]	0.555	0.605	0.775	0.648	0.496	0.530	0.602
Raw [25]	*0.761*	*0.628*	*0.775*	*0.648*	*0.655*	*0.620*	*0.681*
Raw	0.766	0.629	0.775	0.648	0.634	0.631	0.681
Gray imageᵇ [25]	0.888	0.772	0.847	0.686	0.652	0.716	0.760
Our method (FC)	**0.854**	**0.810**	**0.854**	**0.656**	0.540	0.591	0.718
Our method (CNN1D)	**0.862**	**0.818**	**0.863**	0.609	0.564	**0.704**	0.737
MetAML [9]	**0.877**	**0.805**	**0.809**	0.644	0.664	**0.703**	0.750

5   Result

Overall, the FS using Ridge Regression method applied in this gene datasets made a drastic increase in accuracy and the time are less consumed compared to the other remain and previous methods. The ultimate results are carefully elaborated as follows. The charts in this part shows the average performance of training sets on the left and other side illustrates the average performance of test sets. Figures 1, 2 and 3 show the accuracy resulted by feature selection methods of 6 dataset including CIR, COL, IBD, OBE, T2D, WT2D, respectively.

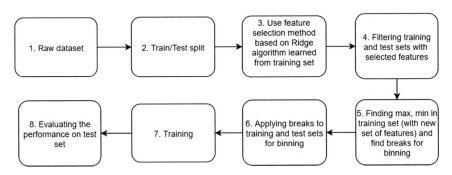

Fig. 1 Steps of applying feature selection on gene family abundance dataset

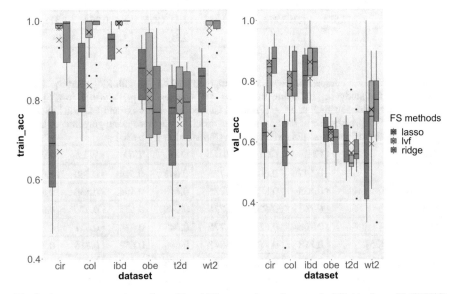

Fig. 2 Accuracy comparison of considered FS approaches using equal width binning with CNN1D

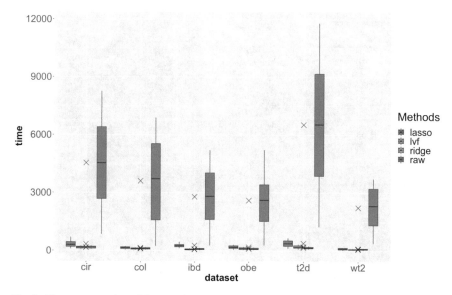

Fig. 3 Time consumption of the experiments using various feature selection methods

- **Feature selection on gene family abundance with binning**

As shown in Fig. 2, while mainly use feature selection based on ridge regression, we also implement Lasso and LVF to compare efficiency between the feature selection methods. Obviously, there is a significant different in mean accuracy of ridge which outperforms the other two in many datasets such as CIR, COL, WT2D. Lasso method seems not to be accurate in terms of preprocessing the datasets when mean of accuracy after training mostly lower than that of the other methods.

- **Reducing Execution time with the proposed method**

It can be observed from Fig. 3 that running time of model training using types of feature selections are significantly less than that of training without any feature selection method intervened. Since less features come into a training architecture, time and memory complexity are highly reduced. Experiments shows that, on every dataset, the time consumed using feature selection are eliminated, and the results are, if having been waiting for hours, cut down to seconds. The following figures also shows that select correct features not only make training consume less time, but also results in more accurate models.

- **Features selected for classifying disease efficiently through experimental results**

Figure 4 exhibits higher accuracy of WT2D as well as CIR, COL and IBD when Ridge regression is applied compared to the result using FC model. The following

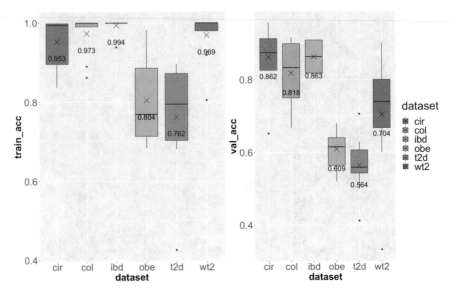

Fig. 4 Final training result on 6 datasets when applying feature selection based on Ridge regression with EQW binning using CNN1D model. CIR, COL, IBD, OBE, T2D, WT2D (from left to right, respectively)

table represents the accuracy values of methods. We will make some comparisons as well as calculate the *p*-values for significance judgement.

Table 2 compares the performance of our methods and state-of-the-art in [9, 25]. Although the work in [25] performed poorly on dimensionality reduction such as PCA and RD_PRO, we obtain promising results with Ridge and equal width binning. Even though, some results for COL and IBD outperforms the results with species abundance using one of the best representations in [25]. Authors in [25] did not provide standard deviation, so we rerun FC on raw data with 10-cross validation for t-test statistics. In comparison between our method using Ridge combined binning and raw data, we compute *p*-value to find the significant improvements (if *p*-value <0.05). For FC model, we achieve two out of six significant results on COL, IBD with *p*-values of 0.0002474, 0.0002474, respectively while we also obtain a *p*-value of 0.05741 on CIR dataset. Using CNN1D, we improve the results with three significant results on CIR, COL, IBD with *p*-values of 0.03942, 0.0002474, 0.0001583, respectively. We also obtain significant results comparing to species abundance on MetAML [9] and gray images from [25] on IBD and comparative results for CIR, COL, WT2.

6   Conclusion

We proposed a way to combine between feature selection approaches and equal width binning to predict diseases on gene families efficiently. On the public benchmark consisting of six family abundance datasets related to five different diseases, we achieve three out of six significant improvement with deep learning algorithms, other datasets can obtain comparative results. The new sets of features after feature selection with only more than one thousand of features can reach to a promising result compared to the original features set including more than one million of features. Furthermore, the execution time for the experiments are improved remarkably. Clearly, the learning with a very number of features consumes a huge amount of time while with a significant smaller number of features, the execution time is acceptable. As shown from our results, we note that gene family can outperform species abundance in predicting some disease such as COL and IBD. A further research should consider and investigate more on this information. With methods of feature selection, we can compare the importance between two these data types. In the future, we research can investigate methods to find an optimal value for the number of features which should be retained. More studies should also explore deeper architecture aiming to obtain better results.

References

1. Handelsman J (2004) Metagenomics: application of genomics to uncultured microorganisms. Microbiol Mol Biol Rev 68:669–684
2. Behjati S, Tarpey PS (2013) What is next generation sequencing? https://doi.org/10.1136/arc hdischild-2013-304340
3. Ehrlich SD (2016) The human gut microbiome impacts health and disease. CR Biol 339(7–8):319–323. https://doi.org/10.1016/j.crvi.2016.04.008 (PMID: 27236827)
4. Truong DT et al (2015) MetaPhlAn2 for enhanced metagenomic taxonomic profiling. Nat Methods 12:902–903
5. NIH HMP Working Group, Peterson J, Garges S et al (2009) The NIH Human Microbiome Project. Genome Res 19:2317–2323. https://doi.org/10.1101/gr.096651.109
6. Fabijanić M, Vlahoviček K (2016) Big data, evolution, and metagenomes: predicting disease from gut microbiota codon usage profiles. In: Carugo O, Eisenhaber F (eds) Data mining techniques for the life sciences. Methods in molecular biology, vol 1415. © Springer Science+Business Media, New York. https://doi.org/10.1007/978-1-4939-3572-7_26
7. Ditzler G et al (2015) Fizzy: feature subset selection for metagenomics. BMC Bioinform 16:358. 10.1186/s12859-015-0793-8
8. Cai L, Wu H, Li D, Zhou K, Zou F (2015) Type 2 diabetes biomarkers of human gut microbiota selected via iterative sure independent screening method. PLoS ONE 10(10):e0140827
9. Pasolli E et al (2016) Machine learning meta-analysis of large metagenomic datasets: tools and biological insights. PLOS Comput Biol. https://doi.org/10.1371/journal.pcbi.1004977
10. Zou H, Hastie T (2005) Regularization and variable selection via the elastic net. J R Stat Soc: Series B (Stat Methodol) 67(2):301–320
11. Hacılar H et al (2020) Inflammatory bowel disease biomarkers of human gut microbiota selected via ensemble feature selection methods

12. Lazar C et al (2012) A survey on filter techniques for feature selection in gene expression microarray analysis. https://doi.org/10.1109/TCBB.2012.33
13. Liu H. Evolving feature selection
14. Statnikov A et al (2013) A comprehensive evaluation of multicategory classification methods for microbiomic data. Microbiome 1(1):11. https://doi.org/10.1186/2049-2618-1-11 (PMID: 24456583)
15. Wagner A et al (1994) Surveys of gene families using polymerase chain reaction: PCR selection and PCR drift. https://doi.org/10.1093/sysbio/43.2.250
16. Nguyen TH, Zucker J (2019) Enhancing metagenome-based disease prediction by unsupervised binning approaches. In: 2019 11th international conference on knowledge and systems engineering (KSE), Da Nang, Vietnam, pp 1–5
17. Qin J et al (2010) A human gut microbial gene catalogue established by metagenomic sequencing. Nature 464(7285):59–65. https://doi.org/10.1038/nature08821 (PMID: 20203603)
18. Qin N et al (2014) Alterations of the human gut microbiome in liver cirrhosis. Nature 513(7516):59–64. https://doi.org/10.1038/nature13568 (PMID: 25079328)
19. Zeller G et al (2014) Potential of fecal microbiota for early-stage detection of colorectal cancer. Mol Syst Biol 10(11):766. https://doi.org/10.15252/msb.20145645.
20. Le Chatelier E et al (2013) Richness of human gut microbiome correlates with metabolic markers. Nature 500(7464):541–546. https://doi.org/10.1038/nature12506 (PMID: 23985870)
21. Qin J et al (2012) A metagenome-wide association study of gut microbiota in type 2 diabetes. Nature 490(7418):55–60. https://doi.org/10.1038/nature11450 (PMID: 23023125)
22. Nguyen TH (2019) Metagenome-based disease classification with deep learning and visualizations based on self-organizing maps. Lecture notes in computer science book series (LNCS), vol 11814. Springer. ISSN: 0302-9743
23. Pasolli E et al (2017) Accessible, curated metagenomic data through experiment hub, pp 1023–1024. ISSN 1548-7105
24. Abubucker S et al (2012) Metabolic reconstruction for metagenomic data and its application to the human microbiome, vol 8, pp e1002-358. ISSN 1553-7358
25. Nguyen TH et al (2019) Disease prediction using synthetic image representations of metagenomic data and convolutional neural networks. In: IEEE Xplore

Mathematical Model Based PID Control of the Raspi Drone

Sanjaa Bold, Batchimeg Sosorbaram, and Sumya Chuluunjav

Abstract This research aiming to study and identify the drone's movement and factors of external influence that interfere with it, and to analyze the ways of keeping a drone's balance position. In order to assure certain improvements, we have developed a homogeneous algorithm, or a model, that would stabilize a drone during flight. The reason of our interest in the drone's stabilization system is following: identification systems require a drone to have steady speed. It is very difficult to run the named system on a drone when its speed, height and location fluctuate. Drone's dynamic stability, height and velocity directly affect the accuracy of detection, identification, comparison and identification systems. Therefore, here was the need for researching the drone's stability and balancing for the purpose of detection systems. Within this framework of studying the drone's dynamic stability, the mentioned algorithm has been developed and further improved according to specific theoretical and practical data. First, we will identify the following four dimensions in 3D space: "thrust", "pitch", "yaw", and "roll". Each of the measurement is expressed in the XYZ axis where their movement direction is recorded as θ, ψ, φ, and identified by the angle change. To do this, we based on the Newton–Euler equation theory. Also, the named theoretical equation was used and improved in this system.

Keywords Theory of the Newton–Euler's and Euler's · Theory of the aerodynamic · Euler's angle · Coordinate system · Acceleration aglne · Matlab module · Raspi drone

S. Bold (✉) · B. Sosorbaram · S. Chuluunjav
Department of Computer Science, University of the Humanities, Ulaanbaatar, Mongolia
e-mail: sanjaa@humanities.mn

B. Sosorbaram
e-mail: chimeg@humanities.mn

S. Chuluunjav
e-mail: sumya@humanities.mn

© The Author(s), under exclusive license to Springer Nature Singapore Pte Ltd. 2021
H. Kim et al. (eds.), *IT Convergence and Security*, Lecture Notes
in Electrical Engineering 712, https://doi.org/10.1007/978-981-15-9354-3_3

1   Introduction

Over the past years, the drone or unmanned aerial vehicle (Unmanned Aerial Vehicle—UAV) research and various projects involving it have expanded significantly (*For the purpose of this research the unmanned aerial vehicle shall be referred to as drone*). Drones are one of the most widely used data collection devices. International scientists and research teams continuously develop various aspects of drones as can be seen from the drone development history. Most importantly, those newly invented algorithms, theories, practices, and experiments show ever improving interaction between a man and drone based on the technical and technological means.

There are currently many algorithms being studied and some are being used for determining how a drone receives a human command and responds to it according to certain algorithms. Bases on advanced algorithms, today drones are widely used in healthcare, communications, delivery and postal service, environmental protection, military and civil defense, entertainment and other fields. As these sectors differ from one another in their utilization of advanced technologies, the drone development also varies greatly depending on each industry. We are researching the drone for the purpose of using in the civil defense. It is being developed for the detection system identification by human body and face proportions, and by other human biometric data. To increase the consistency and accuracy of image processing it is important to ensure the drone's relative stability.

A drone is an effective device which allows to detect objects, regardless of a distance, through use of high resolution cameras. However, a drone flight can be directly impacted by external and internal environment and weather conditions. This is one of disadvantages of a drone. Another disadvantage is relatively short time period of its flights. Moreover, such weather condition like wind is an overpowering factor. Depending on a speed of wind, the timing of the drone operation varies. In windy conditions, energy consumption of a drone motor varies notably for maintaining its balance during flight, off-taking and gaining altitude depending on many factors. In some cases, the drone may get off the drone's control zone and lost due to wind. Therefore, in recent years, the drone research and development has been continuously improving its performance for the purpose of various industries.

For instance:

- location identifier
- return to the first point command
- rotate around one point
- warning signal (to inform the approaching end of the control zone)
- M2M connection are being intensively studied.

In order to improve the accuracy of detection, we worked on improving the drone's capacity to keep flight balance and quickly switch to a steady balances position. The PID management system is very effective in maintaining the drone's flight stability and in using various algorithms in detection systems, human face identification and vehicle license plate number detection. The principle of operating the detection system is to make an object's image during a flight and store it in the database. Depending on the algorithm utilized, the data will be processed to detect a face or sign. However, if the image resolution is inadequate the detection algorithm cannot provide a good result. Therefore, we have developed some technical specifications and mathematical model or algorithm to improve the above. Firstly, we attempted to improve the performance of drone cameras which gave adequately good test results. But our experiment was performed on a windless day. The next experiment in windy condition indicated are very poor system performance due to the wind impact. External environment makes a flight relatively unstable. So, we have to reconsider the improvement. By way of experimenting and analyzing all unsatisfactory test results, we developed an additional algorithm for a wind speedometer and anti-wind power. In case a drone is caught up by wind or flies in wrong direction due to wind power this algorithm allows to launch another algorithm (PID) which ensures the drone's better stability. We reconsidered the main motion equation which studies the drone motion law. The movement equation is determined under the Newton and Euler theory. Also, the motion equation was calculated based on the 3-dimensional space with direction, trajectory and vector which is so-called cylindrical coordinate system. The drone's tilt and angle change due to wind power and other similar external factors were determined by certain mathematical methodology. The practical calculations and equations of the cylindrical coordinate system traffic will be discussed in detail in the next section where they are based on the dependency of primary and secondary calculations system movements.

However, motion data have been processed by the PID system, and the results of some mathematical models expressed in the cylindrical coordinates system have been described in graphic representation. The data were processed under the maximum accuracy method pertaining to the control system and the results of such processing were prepared.

2 Movement Modeling

2.1 Basic Movement Parameters

To determine the mutual dependency of key parameters, the PID control system was expanded and tested. The value of projection of drone location in three-dimensional space is relatively irregular. In order to boosting the capacity to maintain a certain level in the movement location, a mathematical algorithm of the movement was designed through reducing movement errors that occur during change of balance and through estimating the PID determinant parameters. For calculating the mathematical model of the drone's motion, the motion has been developed based on the cylindrical coordinate system.

For studying changes in drone's movement locations a mathematical model of the motion equation was developed based on the coefficient, parameter value, and their

Fig. 1 The 3-dimensional
space with direction,
trajectory and vector

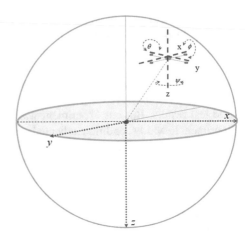

correlation of the PID determinant. The model is characterized by a number of laws, among which the most important laws are the Newton–Euler's and Euler's angle tilt equation laws which require considerable calculation efforts (see Fig. 1) [1].

To determine the mutual dependency of key parameters, the PID control system was expanded and tested. The value of projection of drone location in three-dimensional space is relatively irregular. In order to boosting the capacity to maintain a certain level in the movement location, a mathematical algorithm of the movement was designed through reducing movement errors that occur during change of balance and through estimating the PID determinant parameters. For calculating the mathematical model of the drone's motion, the motion has been developed based on the cylindrical coordinate system.

For studying changes in drone's movement locations a mathematical model of the motion equation was developed based on the coefficient, parameter value, and their correlation of the PID determinant. The model is characterized by a number of laws, among which the most important laws are the Newton–Euler's and Euler's angle tilt equation laws which require considerable calculation efforts (see Fig. 1) [1].

2.2 Dynamic Movement Modeling

For determining the drone's movement, a speed of rotation of four propellers are increased or reduced leading to desired steering. And for designing a dynamic motion model, it is essential to analyses the impact of air resistance and other external forces. Thus, the motion model has been developed correlating to several values in the three-dimensional cylindrical coordinate system. For accomplishing this, the Newton law and Newton–Euler law were studied, and certain improvements were made [1, 2]. For determining changes of the drone's movement, the initial and secondary calculating

Fig. 2 The position of the
first and second coordination
system

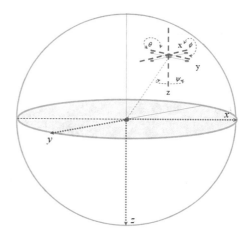

systems was selected, and the movement drifts have been analyzed by reviewing the
changes in values within these calculating systems (Fig. 2).

The parameters determining the direction of the movement.
For it:

- $E = \{E_x, E_y, E_z\}$ initial calculating systems
- $B = \{B_x, B_y, B_z\}$ secondary calculating systems
- $\xi = [x, y, z]^T$ position of calculating system
- $\alpha_{all} = [\varphi, \theta, \psi]^T$ angles on the coordinate system.

The drone's movement direction is represented in the cylindrical coordinate
system as values of $\xi = [x, y, z]^T$ points, they are placed in the original and
secondary coordinate systems. Then, changes of the drone's initial and secondary
locations are represented by the Euler's angle of Roll, Pitch, and Yaw.

When the drone's motion is calculated in the counting system, and the change of
points and angles of each of x, y, z axes are defined by the orthogonal matrix method.

Rotation matrix size is 3×3 [3, 4]. In other words, the rotation matrix is expressed
in three-dimensional metrics: Rot (x, φ), Rot (y, θ), Rot (z, ψ) (Fig. 3).

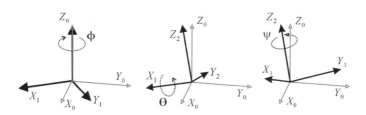

Fig. 3 The position of the rotation coordination system

The drone movement and divergence of a rotation angle in the calculation system which is the ratio between the drone angle and velocity ω, are expressed as equation through Eq. 1.

$$R_s(\omega) = R_x(\varphi) * R_y(\theta) * R_z(\psi) \tag{1}$$

$$R_x(\varphi) = \begin{bmatrix} 1 & 0 & 0 \\ 0 & \cos\varphi & \sin\varphi \\ 0 & -\sin\varphi & \cos\varphi \end{bmatrix} \tag{2}$$

$$R_y(\theta) = \begin{bmatrix} \cos\theta & 0 & \sin\theta \\ 0 & 1 & 0 \\ -\sin\theta & 0 & \cos\theta \end{bmatrix} \tag{3}$$

$$R_y(\psi) = \begin{bmatrix} \cos\psi & \sin\psi & 0 \\ -\sin\psi & \cos\psi & 0 \\ 0 & 0 & 1 \end{bmatrix} \tag{4}$$

Equations 2–4 calculation the rotation in the coordinate system based on equations [5–7].

$$
\begin{aligned}
R_s(\omega) &= R_x(\varphi) * R_y(\theta) * R_z(\psi) \\
&= \begin{bmatrix} 1 & 0 & 0 \\ 0 & \cos\varphi & \sin\varphi \\ 0 & -\sin\varphi & \cos\varphi \end{bmatrix} * \begin{bmatrix} \cos\theta & 0 & \sin\theta \\ 0 & 1 & 0 \\ -\sin\theta & 0 & \cos\theta \end{bmatrix} * \begin{bmatrix} \cos\psi & \sin\psi & 0 \\ -\sin\psi & \cos\psi & 0 \\ 0 & 0 & 1 \end{bmatrix} \\
&= \begin{bmatrix} \cos\psi\cos\theta & \cos\psi\sin\theta\sin\varphi - \sin\psi\cos\varphi & \cos\psi\sin\theta\cos\varphi + \sin\psi\sin\varphi \\ -\sin\psi\cos\theta & \sin\psi\sin\theta\sin\varphi - \sin\psi\cos\varphi & \cos\psi\sin\theta\cos\varphi + \sin\psi\sin\varphi \\ -\sin\theta & \cos\theta\sin\varphi & \cos\theta\sin\varphi \end{bmatrix}
\end{aligned} \tag{5}
$$

To simplify the cos and sin function to make summary of the equation. Therefore, the equation is expressed in symbol A and B.

For it:

$$A_\psi = \cos\psi, \ A_\varphi = \cos\varphi, \ A_\theta = \cos\theta$$
$$B_\psi = \sin\psi, \ B_\varphi = \sin\varphi, \ B_\theta = \sin\theta \tag{6}$$

The equation is dependent on Euler's angle and coordinate system.

$$R_s(\omega) = \begin{bmatrix} A\psi A\theta & A\psi B\theta B\varphi - B\psi A\varphi & A\psi B\theta A\varphi - B\psi B\varphi \\ -B\psi A\theta & B\psi B\theta B\varphi - B\psi A\varphi & A\psi B\theta A\varphi - B\psi B\varphi \\ -B\theta & A\theta B\varphi & A\theta B\varphi \end{bmatrix} \tag{7}$$

The equation of this rotation matrix equation is very important motion equation for drone movement.

2.3 Drone's Aerial Motion Modeling

The main objective of the drone's system for dynamics control is to maintain a motion balance and inertial midpoint, and support returning to the original position after motion diversion. Therefore, it is possible to compare the trajectories, which indicate motion transitions, with help of the two coordinate system for every drone model flying in the three-dimensional space. The motion model is located at relatively fixed points within the initial coordinate system, and these coordinate points would loose an equilibrium position during motion transitioning due to external forces.

Diversion of coordinate points in the 2nd coordinate system is projected over the points in 1st coordinate system, and the changes of coordinate points are recorded (Fig. 4). Diversion of coordinate points caused by air resistance are projected over other axes and recorded as well, thus, a motion equation is calculated (Fig. 4).

We calculate the point in the first and second coordinate of the system by this equation.

$$m\ddot{x} = (\cos\phi \sin\theta \cos\psi + \sin\phi \sin\psi)S_1$$
$$m\ddot{y} = (\cos\phi \sin\theta \sin\psi - \sin\phi \sin\psi)S_1$$
$$\ddot{z}m = -mg + (\cos\phi \cos\theta)S_1 \tag{8}$$

- S_1—Value
- m—mass
- g—gravitational acceleration

Fig. 4 The coordinate system on parameters on drone stabilization

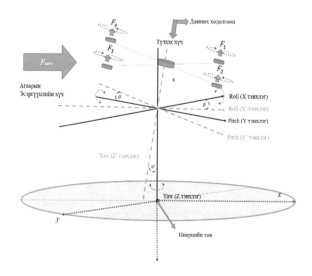

Then, the angle changes are analyzed taking into consideration changes in coordinate points on the 1st and 2nd coordinate systems. The following five conditions will be considerate when calculating the motion model. These parameters are S_1, S_2, S_3, S_4 and the angular speed Eqs. (9)–(10).

$$I_x\ddot{\varphi} = \dot{\theta}\dot{\psi}(I_y - I_z) - j_r\theta\omega + lS_2$$

$$I_y\ddot{\theta} = \dot{\varphi}\dot{\psi}(I_z - I_x) - j_r\varphi\omega + lS_3$$

$$I_x\ddot{\psi} = \dot{\theta}\dot{\varphi}(I_x - I_y) - j_r\psi\omega + lS_4$$

$$S_1 = F_{PRF}, \; S_2 = \tau_\varphi, \; S_3 = \tau_\theta, \; S_4 = \tau_\psi \tag{9}$$

These four identified variables indicate that the drone's displacement and angular changes occur at the coordinates defined at the drone's relocation. The impact of air forces is directly dependent on the dynamic changes of the coordinate system. Depending on the air forces impact, the amount of force produced by a drone's engine varies. From this perspective, the force produced by a drone's engine will vary also due to tilt angle and the transition trajectory. The size of engine's force during transitioning to the drone's balance position was determined by the equation below based on the Newton–Euler method.

$$F_{sum} = F_{motor} + F_{PRF} + F_{GF} \tag{10}$$

In case force of air resistance is determined by the laws of aerodynamics, gravity shall be determined by the law of attraction. To calculate the power generated by the drone's motor, the amount of push-up force should be added because of the law of physics of gravity. Therefore, the amount of engine's force is expressed by the following equation [11, 12]:

$$F_{motor} = R_s(\omega)\left(\sum_{i=1}^{4} F_i - \sum_{i=1}^{4} D_i\right) \tag{11}$$

D_i—drag force

$R_s(\omega)$—rotation matrix and angle acceleration

The air resistance force is calculated by this equation.

$$F_{PRF} = \frac{1}{2}pAC(U^B)^2 \tag{12}$$

– C—air resistance force coefficient

The value of all coordinate axis depends on the coefficient of the air resistance force.

$$C = \text{diag}[C_x, C_y, C_z] \tag{13}$$

At now calculate at now calculate Eqs. 9–11 equations. The defined basic equation for the drone rotation dynamic control system.

$$m\ddot{x} = \left[(\cos\phi \sin\theta \cos\psi + \sin\phi \sin\psi) \sum_{i=1}^{4} F_i - \sum_{i=1}^{4} D_{xi} - \frac{1}{2} pAC(U^B)^2 \right]$$

$$m\ddot{y} = \left[(\cos\phi \sin\theta \sin\psi - \sin\phi \sin\psi) \sum_{i=1}^{4} F_i - \sum_{i=1}^{4} D_{yi} - \frac{1}{2} pAC(U^B)^2 \right]$$

$$\ddot{z}m = \left[-mg + (\cos\phi \cos\theta) \sum_{i=1}^{4} F_i - \sum_{i=1}^{4} D_{yi} - \frac{1}{2} pAC(U^B)^2 \right] \tag{14}$$

Equation 14 allows the coordinate system to calculate the orientation and direction of the drone.

3 Proceeding and Outcomes

A mathematical model of dynamic management system has been developed and some calculations have been performed at a certain level. The data processing and analysis of the above equation has performed based on the MatLab software where some outcomes have been reported. Simulation was performed where the coordinate's axis point was considered to be switching in certain t time intervals, and the following results have been obtained.

The results graph below shows time periods consumed for correcting the deviations in coordinates of the experimental model's axis and returning to the initial position, and other indicators. The fact that the drone could not recover to its first, or equilibrium, position within the time periods as we expected was due to other factors. The green graph indicates the expected time periods for correcting the deviations and transitioning to the initial position (Fig. 5). The red graph shows the actual time periods that were measured during the tests. The results of our experiments specify that the mathematical model, or algorithm, that we created, makes it possible to transition to the previous, balance position with precision of 1.5 s depending on the deviated axis.

On the other hand, analysis of external human influences produced the results shown on the graph below. As can be seen on this graph, during off-taking and gaining altitude a drone makes an arc due to air resistance. Depending on a size (wider or

Fig. 5 Result of the drone's stabilization

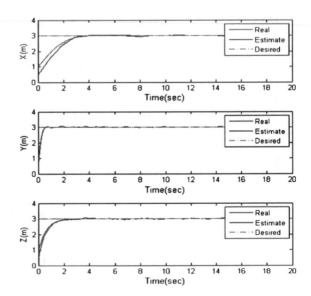

narrower) of the arch's shape, the time periods consumed for switching to a balanced position, vary. A size of the arc also proportionately depends on magnitude of external air force. The results of our experiments proof relatively good specifications of the model we have developed.

4   Conclusion

This study analyses the drone's dynamic motion equation. For developing a mathematical model, we reviewed a significant amount of research materials and textbooks and found out that within the drone development industry the dynamics and modelling a drone's motion model is the field that requires arduous efforts and preparing demanding equations. The results of our experiments show that there have been a number of errors relative to our interpretation of correlation between the computations and data analysis. There were slight delays in certain units during data processing. In contrast to some similar research and development works, our experiments showed more advantages over the time of balance position recovery. The research work we have developed was close to the estimated time of errors in testing and calculations.

References

1. Bold S (2014) Autonomous vision-based moving object detection for unmanned aerial vehicle. Scopus J (Korea)
2. Chingozha T, Nyandoro O (2014) Adaptive sliding backstepping control of quadrotor UAV attitude. In: IFAC 2014 Conference, South Africa, Aug 2014
3. Colmenares-Vazquez J, Marchand N, Garcia PC, Gomez-Balderas J-E (2017) An intermediary quaternion-based control for trajectory following using a quadrotor. In: IROS 2017
4. Bold S, Sosorbaram B (2017) Smart license plate recognition using optical character recognition based on the multicopter. IJRTCC 5(9)
5. Benić Z, Piljek P, Kotarski D (2016) Mathematical modelling of unmanned aerial vehicles with four rotors. Interdiscip Descr Complex Syst 14(1):88–100
6. Rodriguez HR, Vega VP, Orta AS, Salazar OG (2014) Robust backstepping control based on integral sliding modes for tracking of quadrotors. J Intell Rob Syst 73(1–4):51–66
7. Gelb A (1974) Applied optimal estimation. MIT Press
8. Nagaraj B, Murugananth N (2010) A comparative study of PID controller tuning using GA, EP, PSO and ACO. In: IEEE international conference on communication control
9. Melo FE, Kienitz KH, Brancalion JFB (2010) Augmentation to the extended Kalman bucy filter for single target tracking. In: Proceedings of the 9th Brazilian conference on dynamics control and their applications, Serra Negra
10. Abo-Hammour ZS, Alsmadi OM, Bataineh SI, Al-Omari MA, Affach N (2011) Continuous genetic algorithms for collision-free cartesian path planning of robot manipulators. Int J Adv Robot Syst 8(6):14–36
11. Ghanbari A, Noorani SM (2011) Optimal trajectory planning design of a crawling gait in a robot using genetic algorithm. Int J Adv Rob Syst 8(1):29–36

Exploration of a Prediction Model of Aggression in Children Using Bayesian Networks

Hyejoo Lee and Euihyun Jung

Abstract Aggression formed in children lasts for their lifetime and it often produces serious antisocial problems. Therefore, a lot of studies have been conducted for this subject in the educational domain, but the studies adopting data mining are relatively few yet. In this paper, the authors adopt Bayesian Networks to find which variables are related to aggression and to evaluate how the variables affect it. Markov Blanket and IMDB method are used to learn a Bayesian Network and to find the relevant variables. In the results, "social withdrawal", "depression", "mobile phone dependency", "grade of Korean", "attention" and "school activity" are extracted as the relevant variables to aggression. Also, this research investigates which variables are most influencing to aggression by changing the probabilities of variables in the learned Bayesian Network.

Keywords Data mining · Bayesian network · Markov blanket · Aggression

1 Introduction

Aggression is intentional behavior, thoughts, or feelings that are done directly or indirectly to harm others physically and psychologically [1]. It associates with a variety of issues such as violence, bullying and youth crime in school life. Also, aggression, formed in children, lasts from childhood to adulthood and it can be developed to antisocial behavior which results in very serious consequences [2]. Therefore, in the educational domain, a lot of studies have been conducted to find the relevant variables of aggression and to explore the relationship between variables [3, 4].

H. Lee
Department of Education, Chung-Ang University, Seoul, Korea
e-mail: ladyzen@cau.ac.kr

E. Jung (✉)
Department of Convergence Software, Anyang University, Anyang City, Kyunggi-do, Korea
e-mail: jung@anyang.ac.kr

© The Author(s), under exclusive license to Springer Nature Singapore Pte Ltd. 2021
H. Kim et al. (eds.), *IT Convergence and Security*, Lecture Notes
in Electrical Engineering 712, https://doi.org/10.1007/978-981-15-9354-3_4

Previous studies have found the various variables related to aggression such as gender, self-esteem, self-control, depression, abusing, parental attachment, phone dependency, etc. [5, 6], but sometimes the results were conflicted because most studies were based on previous studies or hypotheses with conventional statistical methods such as structural equation model analysis. This conventional statistical approach is good to make a solid theoretical model, but the model has to be validated with the only limited variables instead of using all of the variables involved. To remedy this, some educational researchers have tried to adopt data mining on big data from the education domain.

In data mining methods, Bayesian Networks (BNs) [7] have attracted researchers in various domains due to their intuitive explainability and simulation ability. A BN is a directed acyclic graph where vertices represent variables and directed edges show probabilistic dependencies between the vertices. In a BN, every edge has a direction from a variable X to a variable Y and it means that X will cause Y with probability. Therefore, once a BN structure is discovered from a dataset, researchers can easily figure out the causal semantics between the variables in the dataset with the BN. In addition to this, since BNs enable researchers to change the probability of the variables, the researchers can conduct various analyses such as causal discovery, prediction, classification, and diagnosis by monitoring how the changes of variables' probabilities influence to other variables. However, BNs are rarely adopted yet in the educational domain because educational researchers have preferred conventional statistic methods and BNs require considerable knowledge of probability.

In this paper, the authors adopt BNs to discover which variables are directly and indirectly related to aggression in children and how the variables affect it. For this, the Korea Children and Youth Planet Survey (KCYPS) data [8] is preprocessed and 39 variables in KCYPS are selected. In order to learn a BN structure and to select the relevant variables, Incremental Association Markov Blanket (IAMB) [9] is used with the dataset. During the interpretation of the BN structure, the authors conclude aggression is highly related to "social withdrawal", "depression", "mobile phone dependency", "grade of Korean", "attention" and "school activity".

The rest of this paper is organized as follows. Section 2 describes a dataset for this study and explains how to learn a BN with the dataset. In Sect. 3, the authors interpret the learned BN to evaluate the cause and effect of variables by changing the probabilities of variables, and Sect. 4 concludes the paper.

2 Learning a Bayesian Network Structure

2.1 Overview of a Dataset

A dataset for this research is originated from the Korean Children and Youth Panel Survey (KCYPS) longitudinal data collected by the National Youth Policy Institute (NYPI) of South Korea [8]. The KCYPS conducted seven follow-up surveys from

2010 to 2016. At the time of the first study (2010), the children were in the 4th-grade classes. For this study, data from the third wave (2012) were analyzed and the children were in the 6th-grade classes of an elementary school.

To make a qualified dataset, a data preprocessing has been done. Originally, each variable has a 4-point Likert scale (strongly yes, likely yes, likely no, and strongly no) or a 5-point Likert scale (strongly yes, likely yes, so so, likely no, and strongly no). The 4-point Likert scale values are mapped to a binary group and the 5-point Likert scale values are mapped to a categorical group for learning BNs. Consequenetly, a total of 1,604 subjects were included after dropping out those which had missing fields and 39 variables were selected. The selected 39 variables are summarized in Table 1.

2.2 Learning Baeysian Networks

In order to discover a BN structure, the authors uses Markov Blanket [9] as a learning algorithm. The Markov Blanket in a BN for node X_i which we denote by $MB(X_i)$ is a set of nodes composed of X_i's parents, its children and parents of its children. Formally the definition of Markov Blanket in a BN, or more general in a graph, is as follows.

$$MB(X_i) = Par(X_i) \cup Ch(X_i) \cup \bigcup_{Y \in Ch(X_i)} Par(Y) \qquad (1)$$

Using Markov Blanket, the authors can figure out the highly related variables to aggression from the discovered BN structure as shown in Fig. 1. From the Markov Blanket of *agg* variable, *with*, *dep*, *atten*, and *phdep* variables are directly related. Besides, *grlang* and *schact* variables are indirectly related. These variables are shown in bold text in Table 1.

Besides Markov Blanket, the authors adopt another tool, Netica [10], to get the conditional probability distribution. Figure 1 shows how the initial BN structure with the probabilities is displayed in Netica. For example, in the *dep* node, there is the "yes" value 7.90 which means P(yes = I feel like depressed) is 0.079.

3 Data Interpretation and Discussion

3.1 Causal Reasoning

In BNs, it is possible to predict from causes to effects by changing the probabilities of causal variables. In Fig. 2, the authors maximize the "yes" value of *with* from 26.8 to 100.0% and the "yes" value of *dep* variable from 7.9 to 100.0%. Then, the "yes"

Table 1 The selected variables for data mining with Bayesian Networks

Variable name	Description
dad	Father's final degree (Middle school, High school, College, University, Graduate school)
health	Health (Good, Bad)
gralang	Grade of Korean (Hig, Middle, Low)
gramath	Grade of Mathematics(High, Middle, Low)
graengl	Grade of English (Hig, Middle, Low)
grasat	Satisfying the grade (Yes, No)
mas	The pursuit of mastering (Yes, No) (2 questions, Chronbach's $\alpha = 0.742$)
life	Life satisfaction (Yes, No) (3 questions, Chronbach's $\alpha = 0.862$)
man	Managing studying time (Yes, No) (4 questions, Chronbach's $\alpha = 0.844$)
act	Control one's action (Yes, No) (5 questions, Chronbach's $\alpha = 0.714$)
with	Social withdrawl (Yes, No) (5 questions, Chronbach's $\alpha = 0.886$)
caneg	Caring—Neglect (Yes, No) (4 questions, Chronbach's $\alpha = 0.814$)
caabu	Caring—Abuse (Yes, No) (4 questions, Chronbach's $\alpha = 0.844$)
accom	Accomplishment is important (Yes, No) (7 questions, Chronbach's $\alpha = 0.902$)
atten	Good at attention (Yes, No) (7 questions, Chronbach's $\alpha = 0.835$)
dep	Depression (Yes, No) (10 questions, Chronbach's $\alpha = 0.922$)
comgame	Preference of computer game in using a computer(Yes, No)
mlaug	Misconduct of laughing at others (Yes, No)
mrej	Misconduct of rejection to others (Yes, No)
mbeat	Misconduct of beaing others (Yes, No)
mthre	Misconduct of threatening others (Yes, No)
damlaug	Damage of being laughing at (Yes, No)
damrej	Damage of being rejection (Yes, No)
dambeat	Damage of being beaten (Yes, No)
mthre	Damage of being threaten (Yes, No)
commu	Want to contribute to the community (Yes, No) (4 questions, Chronbach's $\alpha = 0.823$)
schact	Goot at school activity (Yes, No) (5 questions, Chronbach's $\alpha = 0.756$)
schrul	Keep the school rules (Yes, No) (5 questions, Chronbach's $\alpha = 0.811$)
schfri	Get along well with friends (Yes, No) (5 questions, Chronbach's $\alpha = 0.708$)
schtea	Get along well with teachers (Yes, No) (5 questions, Chronbach's $\alpha = 0.891$)
multi	multicultural acceptance (Yes, No) (5 questions, Chronbach's $\alpha = 0.868$)
local	Aware of local society (Yes, No) (6 questions, Chronbach's $\alpha = 0.748$)
phdep	Dependency on phones (Yes, No) (7 questions, Chronbach's $\alpha = 0.890$)

(continued)

Table 1 (continued)

Variable name	Description
papeknow	Parent's knowing of peers (Yes, No)
papemeet	Parent meets peers (Yes, No)
papelike	Parent likes peers (Yes, No)
date	Date with heterosexual friends (Yes, No)
fandom	Activity of fandom (Yes, No)
agg	Aggression (Yes, No) (6 questions, Chronbach's $\alpha = 0.816$)

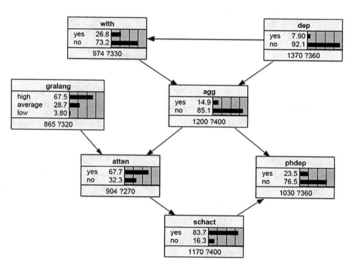

Fig. 1 The initial BN structure composed of variables and their relations. Each variable has the probability

value of *agg* rises up from 14.9 to 58.9%. It means if a child is depressed and feels like social withdrawal, the child will be quite likely to be aggressive.

This kind of experiment enables researchers to observe and compare various cases to determine how much a child is likely to be aggressive by changing the probabilities of the causal variables. In Table 2, the authors summarized the four cases of maximizing the probabilities of the causal variables.

First, when the "yes" of *with* variable and the "yes" of *dep* variable are set 100%, the "yes" of *agg* rises from 14.9 to 58.9%. Second, when the authors change the "no" of *with* and *dep* to be 100%, the "yes" of *agg* goes down from 14.9 to 9.3%. Between two variables, the *dep* variable affected greater than the *with* variable because the probabilities of *agg* is bigger when *dep* is "yes", but the opposite is not. In conclusion, if parents want their children not to be aggressive, parents should keep their youths from getting depressed.

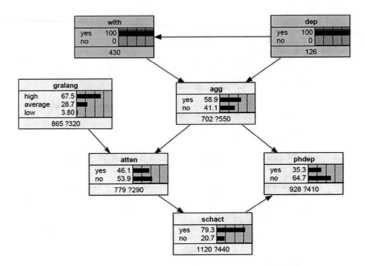

Fig. 2 When the probabilities of the causal variables are set to be changed, the target variable's probability becomes to be changed

Table 2 Changes of causal variables' probabilities and their effects	*with*	*dep*	*agg* (yes)	*agg* (no)
	Yes = 26.8, No = 73.2 (initial values)	Yes = 9.9, No = 92.1	14.9	85.1
	Yes = 100.0	**Yes = 100.0**	**58.9**	41.1
	No = 100.0	**Yes = 100.0**	**50.0**	50.0
	Yes = 100.0	No = 100.0	18.2	81.8
	No = 100.0	No = 100.0	9.3	90.7

3.2 Evidential Reasoning

BNs also enable researchers to perform evidential reasoning. Evidential reasoning is useful to determine the likelihood value of the target variable when the related evidence is given.

In Fig. 3, when the "yes" of *atten* is set from 67.7 to 0.0% and the "yes" of *phdep* is set from 23.5 to 100.0%, *agg* variable's probability rises up from 14.9 to 52.9%. It is about three times higher probability than before. It means if a child is watched not to be good at attention and to have mobile phone dependency, the risk of being aggressive can be three times higher than a normal child. Table 3 shows how the target variable's probability is changed when the evidence is given.

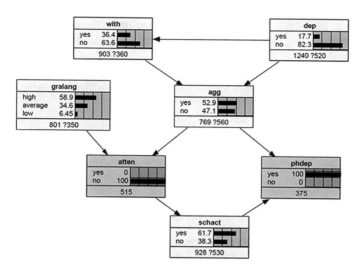

Fig. 3 When pieces of evidence are given, the BN structure can determine the probability of the target variable

Table 3 Changes of the *atten* and the *phdep* variables' value and their effects

atten	*phdep*	*agg* (yes)	*agg* (no)
Yes = 67.7, No = 32.3 (initial values)	Yes = 23.5, No = 76.5	14.9	85.1
Yes = 100.0	Yes = 100.0	12.7	87.3
Yes = 100.0	No = 100.0	3.9	96.1
No = 100.0	**Yes = 100.0**	**52.9**	**47.1**
No = 100.0	No = 100.0	26.2	73.8

4 Conclusion

Aggression formed in children lasts from childhood to adulthood and it often produces serious antisocial problems. Therefore, educational researchers have conducted a lot of studies to figure out which variables are the cause of aggression and to explore the effects among variables. However, in the previous research, it is difficult to predict which and how the multiple causal variables influence aggression.

The research tries to find which variables are in charge of aggression and to evaluate how the variables influence. To do this, a BN structure is learned with IAMB and Markov Blanket. As a result, "social withdrawal", "depression", "mobile phone dependency", "grade of Korean", "attention" and "school activity" are found as the relevant variables to aggression. From causal reasoning, the authors conclude that depression has much stronger effects on aggression than social withdrawal. Also, the authors observe attention is tightly related to aggression from the evidential reasoning. In the future, the authors are going to examine the causal effects of other variables in detail and extend the research to find other meaningful features in the educational domain.

Acknowledgements This work was supported by the Ministry of Education of the Republic of Korea and the National Research Foundation of Korea (NRF-2018S1A5B5A07072578)

References

1. McEvoy MA, Estrem TL, Rodriguez MC, Olson ML (2003) Assessing relational and physical aggression among preschool children: intermethod agreement. Top Early Child Spec Educ 23(2):51–61
2. Kupersmidt JB, Coie JD (1990) Preadolescent peer status, aggression, and school adjustment as predictors of externalizing problems in adolescence. Child Dev 61(5):1350–1362
3. Hipwell A, Keenan K, Kasza K, Loeber R, Stouthamer-Loeber M, Bean T (2008) Reciprocal influences between girls' conduct problems and depression, and parental punishment and warmth: a six year prospective analysis. J Abnorm Child Psychol 36(5):663–677
4. Willoughby T, Adachi PJ, Good M (2012) A longitudinal study of the association between violent video game play and aggression among adolescents. Dev Psychol 48(4):1044
5. Ooi YP, Ang RP, Fung DS, Wong G, Cai Y (2006) The impact of parent–child attachment on aggression, social stress and self-esteem. Sch Psychol Int 27(5):552–566
6. Washburn JJ, McMahon SD, King CA, Reinecke MA, Silver C (2004) Narcissistic features in young adolescents: relations to aggression and internalizing symptoms. J Youth Adolesc 33(3):247–260
7. Jensen FV (1996) An introduction to Bayesian networks. UCL Press, London
8. National Youth Policy Institute (2010) The 2010 Korean children and youth panel survey project report. Seoul, Korea
9. Tsamardinos I, Aliferis CF, Statnikov AR, Statnikov E (2003) Algorithms for large scale Markov blanket discovery. In: FLAIRS conference, vol 2, pp 376–380
10. Norsys Software Corporation (2020) Netica is a trademarks of Norsys software Corporation. https://www.norsys.com/netica.html. Accessed 30 Jan 2020

Deep Learning-Based Logging Recommendation Using Merged Code Representation

Suin Lee, Youngseok Lee, Chan-Gun Lee, and Honguk Woo

Abstract When developing a large scale software product, it is essential to share a common set of structural coding guidelines and standards among the project team members. In this paper, we propose *MergeLogging*, a deep learning-based merged network using various code representations for automated logging decisions or other tasks. *MergeLogging* archives the enhanced recommendation ability that utilizes orthogonal code features from code representations. Our case study with three open-source project datasets demonstrates that logging accuracy can reach as high as 93%.

Keywords Logging recommendation · Code embedding · Deep learning

1 Introduction

For software quality management with deep learning, low-dimensional vector representation utilization can work as a solution, referred to as code embedding which automatically summarizes source code in a fixed-size vector [1]. One main challenge of code embedding is how to capture features of source code. Several code embedding approaches exploit structural path patterns of source code. These patterns are

S. Lee · H. Woo (✉)
Department of Computer Science and Engineering, Sungkyunkwan University, Seoul, Korea
e-mail: hwoo@skku.edu

S. Lee
e-mail: sheen1963@skku.edu

Y. Lee
Department of Electrical and Computer Engineering, Sungkyunkwan University, Seoul, Korea
e-mail: yslee.gs@gmail.com

C.-G. Lee
Department of Computer Science and Engineering, Chung-Ang University, Seoul, Korea
e-mail: cglee@cau.ac.kr

© The Author(s), under exclusive license to Springer Nature Singapore Pte Ltd. 2021
H. Kim et al. (eds.), *IT Convergence and Security*, Lecture Notes
in Electrical Engineering 712, https://doi.org/10.1007/978-981-15-9354-3_5

represented in the form of an abstract syntax tree (AST) that is built via code translation and compiling processes. For instance, code2seq proposed by Uri et al. embeds a random set of compositional paths over AST representations of source code in its deep learning-based prediction model [2].

In this paper, we concentrate on the robustness of code embedding by exploiting different representations of source code. Specifically, we employ the merged deep neural network which intends for integrating all the features extracted from different code representations including original code snippets and structural data from AST. This merged network allows us to explore the orthogonality of code features extracted from text-based code embedding as well as path-based code embedding. We utilize this merge network for automated logging decisions or other recommendation tasks [3]. The main contributions of this paper are summarized as follows:

(1) We propose a deep learning-based logging recommendation model that utilizes different code embedding schemes with a merged deep neural network.
(2) We demonstrate the performance improvement of logging recommendation for several open source datasets. We also make our datasets accessible on Github [4], which have been preprocessed for machine learning cases with easy to use.

2 Proposed Approach

2.1 MergeLogging Model

Figure 1 illustrates the architecture of *MergeLogging*. The process of the model consists of four steps: (1) *Code embedding*, a code snippet is fed to each code embedding network, and is embedded to a sequence of fixed-size vectors as an input document. (2) *Feature extraction*, orthogonal code features are extracted from each document via attentive LSTM networks. (3) *Feature merging*, a merge network combines the code features represented in different dimensions into the same dimension. (4) *Task prediction*, the softmax function stacked on top of *MergeLogging* which is explained in the steps (1)–(3), produces its output as prediction (recommendation).

Text-based Code Embedding Network. A code snippet is embedded as a document with a series of code word vectors using word2vec [5]. This word-wise model

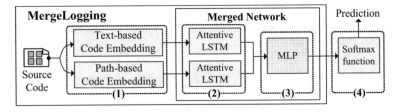

Fig. 1 *MergeLogging* architecture

is trained to reconstruct contexts of word-level tokens in code, similar to what is commonly done for natural language processing.

Path-based Code Embedding Network. Code2seq proposed to use random path sequences extracted via ASTs internally constructed [2]. A code snippet is parsed by Java parser (i.e., Jpredict) to construct a respective AST. Then, an AST is traversed in a way that syntactic paths from a leaf node to another leaf node via the root node are extracted. After extracting multiple random paths, each path is encoded by the AST encoder so as to be combined on the vector representation. Finally, the decoder predicts the output sequence of tokenized method names (i.e., add|difference) with attention mechanism. Having this method of name-labeled supervised learning, we can obtain the intermediate output of embedding layer with low-dimensional vectors.

Attentive LSTM Network. For a document with code embedding vectors $\mathbf{E} = [e_1, \ldots, e_l]$, an l-length input vector sequence, an attentive LSTM function computes an o-size output feature vector $\mathbf{f} = [f_1, \ldots, f_o]$:

$$\mathbf{f} = \text{LSTM}_{\text{att}}(\mathbf{E}) \tag{1}$$

The attentive LSTM function calculates the intermediate output vector sequence $\mathbf{F} = [\mathbf{f}_1, \ldots, \mathbf{f}_l]$, the hidden state vector $\mathbf{h}$, and the attention weight vector $\mathbf{a}$ by iterating the following equations for $t = 1, \ldots, l$ and $t' = 1, \ldots, i$:

$$\mathbf{h}_{t,t'} = \mathcal{H}\left(W_{\text{eh}}\mathbf{e}_t + W_{\text{hh}}\mathbf{h}_{t-1,t'} + \mathbf{b}_t\right), \quad \mathbf{m}_t = \sigma\left(W_a\mathbf{h}_{t,t'} + \mathbf{b}_a\right) \tag{2}$$

$$\mathbf{a}_t = \text{SOFTMAX}(\mathbf{m}_t), \quad \mathbf{f}_t = \sum_{t'} a_{t,t'}\mathbf{e}_t \tag{3}$$

Here, $\mathcal{H}(\cdot)$ is the hidden layer activate function, $\sigma(\cdot)$ is the sigmoid function, SOFTMAX$(\cdot)$ is the sofxmax function, W_{eh} is the input-hidden weight matrix, W_{hh} is the hidden layer weight matrix, and $\mathbf{b}$ is the bias vector. In addition, an output feature vector $\mathbf{f}$ is obtain by concatenating all elements of the intermediate output vector sequence $\mathbf{F}$.

2.2   Model Implementation

We set the vector size of word2vec and code2seq used in text- and path-based code embedding to 200 and 320 respectively. The input vector sequence length and the output vector size of each attentive LSTM network are set to 200 and 25,600 respectively. In addition, the input and output vector size of deep MLP network with three layers and a Leaky ReLU activate function are 51200 and 2 respectively. The value of learning rate is set to 0.0004.

Table 1 Logging recommendation accuracy

Dataset	Texted-based	Path-based	*MergeLogging*
Hadoop	87.0	86.8	**88.3**
Elastic Search	91.6	92.5	**93.3**
Spring Framework	88.7	88.9	**90.2**

3   Evaluation

In this section, we evaluate the performance of our proposed model, compared to those of the text- and path-based (code2seq) code embedding models. We implemented the models using Keras framework, and executed tests on a system with an Intel CPU i9-9940X processor 3.30 GHz (14cores), 128G of RAM, and NVIDIA Geforce GTX 2080 SLI. All tests were performed with tenfold cross-validation sets. We utilized famous open source projects in Java (i.e., Hadoop, Elastic Search, and Spring Framework) as datasets that are obtained by labeling each method with logging recommendation. The datasets can be accessed on Github [4] where the number of samples in each dataset is 20336, 14310, and 26372.

Table 1 provides an overall performance comparison for logging recommendation. The text- and path-based models turn out to show the same level of accuracy, while they are intended to learn different features of source code; the text- and path-based models focus on keywords and code structures respectively. *MergeLogging* outperforms these individual models, obtaining a gain more than 1–2%. It is noticed that *MergeLogging* combines the advantages of text- and path-based code embedding via the merge network, hence resulting in the highest accuracy of the comparison group.

4   Conclusion

In this paper, we presented the deep learning-based logging recommendation model that exploits the text-based as well as path-based code embedding. The proposed *MergeLogging* model showed a competitive logging recommendation accuracy across the datasets of three open source projects. While our implementation has adopted only embedding networks for texts and paths, the model structure is extensible to include another network which handles a different code representation, e.g., binary code.

Acknowledgements This research was supported by Basic Science Research Program through the National Research Foundation of Korea (NRF) funded by the Ministry of Science, ICT and Future Planning (NRF-2017R1E1A1A01075803, NRF-2018R1D1A1A02086102, NRF-2020R1A2C2009809).

References

1. Gao Z, Jayasundara V, Jiang L, Xia X, Lo D, Grundy J (2019) SmartEmbed: a tool for clone and bug detection in smart contracts through structural code embedding. In: 35th international conference on software maintenance and evolution. IEEE, Cleveland, OH, pp 394–397
2. Alon U, Brody S, Levy O, Yahav E (2019) code2seq: generating sequences from structured representations of code. In: 7th international conference on learning representations, ICLR 2019, New Orleans, LA, pp 1–22
3. Kiczales G, Hilsdale E, Kersten M, Palm J, Griswold GG (2001) An overview of AspectJ. Springer, Heidelberg
4. Datasets for Java Logging Recommendations (2020) https://github.com/dooinee/mergeLogging. Accessed 20 Feb 2020
5. Mikolov T, Chen K, Corrado GS, Dean J (2013) Efficient estimation of word representations in vector space. In: 1st international conference on learning representations, ICLR 2013, Scottsdale, AZ, pp 1–12

Part II
Communication and Signal Processing

Analysis of the Influence of the Channel Model on the Optimum Switching Points Determination in a Hybrid Adaptive Modulation and FEC System

E. S. Diogo and J. M. C. Brito

Abstract In recent years, adaptive modulation and adaptive FEC techniques have received considerable attention for improving the performance of wireless networks. The analysis of switching points is a key factor in maximizing the performance of systems using these techniques. In this paper, it is analyzed the ideal switching points of an adaptive hybrid approach that combines adaptive modulation and adaptive FEC techniques, considering a Nakagami-m fading channel. The criteria of maximum throughput and power metric is used to compute the optimum switching points. Also, we analyzed the influence of the channel model in determining the ideal switching points; for this purpose, the parameter m of the Nakagami-m model was varied. It showed up with the results achieved that the best switching points depend on: the channel model and the selected performance criterion.

Keywords Hybrid adaptive modulation and FEC · Throughput criterion · Power metric criterion · Nakagami-m fading · Optimum switching points

1 Introduction

The emerging demand for new types of services in wireless networks points to the design of increasingly intelligent and agile communications systems capable of providing higher data rates and more efficient and flexible use of the spectrum. The techniques of adaptive modulation and adaptive FEC are adequate alternatives to reach the mentioned requirements [1–4]. They are designed to monitor channel variations, changing the modulation scheme and the coding rate to increase efficiency, with high rates of information transmission under favorable channel conditions and

E. S. Diogo (✉) · J. M. C. Brito
Instituto Nacional de Telecomunicações (INATEL), Santa Rita do Sapucaí, MG, Brazil
e-mail: elvira.salvador@mtel.inatel.br

J. M. C. Brito
e-mail: brito@inatel.br

© The Author(s), under exclusive license to Springer Nature Singapore Pte Ltd. 2021
H. Kim et al. (eds.), *IT Convergence and Security*, Lecture Notes in Electrical Engineering 712, https://doi.org/10.1007/978-981-15-9354-3_6

reducing the rate of information transmission in response to the degradation of the channel [1, 2].

In adaptive modulation systems, the order of the modulation (number of symbols in the constellation of the modulation) is adjusted according to the channel quality. By reducing the order of the modulation, the Bit Error Rate (BER) and the data rate decrease. In an adaptive Forward Error Correction (FEC) system, the number of redundancy bits or symbols are adjusted in the transmitter according to the channel conditions. By increasing the number of redundancy bits, the error-correcting capacity of the code increases, reducing the probability that a packet contains errors that can not be automatically corrected [1–3].

Finally, in a hybrid adaptive system, both the number of redundancy bits and order of modulation are adjusted according to the channel conditions. An important aspect in adaptive systems is to define the ideal switching points between neighboring modulations or codes according to some desired performance criteria. In the analysis presented in this paper, two modulations have been considered to be neighbors if the order of one modulation is $M = 2^x$, and the order of the other one is $M' = 2^{x-1}$ or $M'= 2^{x+1}$. Similarly, two error correction codes have been considered to be neighbors if the error correction capacity of one of them is equal to t bits and the capacity of the other one is equal to $(t + 1)$ or $(t - 1)$ bits.

The analysis presented in [5] shows that hybrid systems have performance equal to or better than systems that use only adaptive modulation or only adaptive error control schemes. However, the analysis made in [5] only applies to some specific types of wireless networks, since it was considered a channel without memory. The performance of adaptive modulation techniques and adaptive FEC for channels with fading were investigated separately in several studies [1, 2, 6–11]. However, the optimum switching points of a hybrid system under fading channels have not yet been analyzed.

The contribution of this paper is to analyze the influence of the channel model in determining the ideal switching points of a hybrid adaptive scheme, which uses adaptive FEC in conjunction with adaptive modulation, considering throughput and power metric as performance parameters. A Nakagami-m fading channel is considered in the analysis.

The remainder of this paper is organized as follows: Sect. 2 presents the channel model; Sect. 3 develops the calculation of the packet error rate (PER) in the channel; the throughput criterion is presented in Sect. 4, and the power metric criterion is presented in Sect. 5; Sect. 6 shows the numerical results, and Sect. 7 presents the main conclusions.

2 Channel Model

In this paper, is considered a Nakagami-m fading channel. The Nakagami-m distribution uses a parameter (m) to describe the degree of fading suffered by the signal propagating in a multipath environment, and it also offers a good fit for urban and indoor multipath propagation [12]. Rayleigh's distribution is a particular case of Nakagami-m's distribution.

When $m = 0.5$, the Nakagami model describes a unilateral Gaussian distribution, corresponding to a scenario with the occurrence of a large paths number. When $m = 1$, the channel behaves as a Rayleigh channel. For values above 1, a one-to-one mapping occurs between the Nakagami parameter and the Ricean factor, causing the Nakagami distribution to approach a Rice distribution [13].

The PDF of the Nakagami-m distribution as a function of the SNR is given by [14]:

$$P_\gamma(\gamma) = \frac{m^m \gamma^m}{\left(\overline{\gamma}\right)^m \Gamma(m)} \exp\left(-\frac{m\gamma}{\overline{\gamma}}\right), \tag{1}$$

where $\overline{\gamma}$ is the average Signal-to-Noise Ratio (SNR) received, and $\Gamma(m)$ is the Gamma function, defined by $\Gamma(m) = \int_0^\infty x^{m-1} e^{-x} dx$.

3 Model to Compute the Packet Error Rate

In this paper, is considered an (n, k) binary linear block code, where k is the number of information bits, and $(n - k)$ is the number of parity bits in the code. Defining the error correction capacity of the code as t, the probability that a block of n bits contains errors that can be automatically corrected is given by the binomial distribution [4]:

$$PER = \sum_{i=0}^{t} \binom{n}{i} P_{BER}^i \times (1 - P_{BER})^{n-i}, \tag{2}$$

where P_{BER} is the average bit error rate in the system. In a hybrid adaptive system, the value of n depends on the current code and the value of P_{BER} depends on the current modulation.

In our analysis, is considered the BCH (Bose, Chaudhuri, and Hocquenghem) block codes. These codes represent cyclic codes and are a generalization of the Hamming codes, allowing for corrections of multiple bit errors. For any positive integer's m ($m \geq 3$) and t ($t < 2^{m-1}$), there exists a binary t-error-correcting BCH code with the following rules [4]:

$$\begin{cases} n = 2^m - 1 \rightarrow \text{ length of the block} \\ (n - k) \leq mt \rightarrow \text{ number of parity bits} \\ d_{min} = 2t + 1 \rightarrow \text{ minimum distance of the code} \end{cases}$$

where m is an arbitrary integer, and d_{min} is the minimum distance of the code.

3.1   Calculation of BER

Following [15], to compute the exact BER, a rectangular *M-QAM* is modeled as two independent *I-ary* and *J-ary* (PAM) pulse amplitude modulations, where $M = I \times J$. The average bit error probability for rectangular *M-QAM* is given by [15]:

$$P_{bn} = \frac{1}{\log_2(I \times J)} \left[\sum_{k=1}^{\log_2 I} P_I(k) \times \sum_{l=1}^{\log_2 J} P_J(l) \right], \tag{3}$$

where $P_I(k)$ is the error probability of the k_{th} bit of the in-phase component, calculated by an *I-ary* PAM modulation, for $k \in \{1, 2, ..., \log_2 I\}$, given by [15]:

$$P_I(k) = \frac{1}{I} \sum_{i=0}^{(1-2^{-k})I-1} \left\{ (-1)^{\frac{i*2^{k-1}}{I}} \times \left[2^{k-1} - \frac{i*2^{k-1}}{I} + \frac{1}{2} \right] \times \right.$$
$$\left. \text{erfc}\left((2i+1)\sqrt{\frac{3\log_2(I*J)*\gamma_b}{I^2+J^2-2}} \right) \right\}, \tag{4}$$

and $P_J(l)$ is the error probability of the l_{th} bit of the quadrature component, calculated by a *J-ary* modulation PAM, for $l \in \{1, 2, ..., \log_2 J\}$, computed by [15]:

$$P_J(l) = \frac{1}{J} \sum_{j=0}^{(1-2^{-l})J-1} \left\{ (-1)^{\left\lfloor \frac{j*2^{l-1}}{J} \right\rfloor} \times \left[2^{l-1} - \left\lfloor \frac{j*2^{l-1}}{J} + \frac{1}{2} \right\rfloor \right] \times \right.$$
$$\left. \text{erfc}\left((2j+1)\sqrt{\frac{3\log_2(I*J)*\gamma_b}{I^2+J^2-2}} \right) \right\}, \tag{5}$$

where γ_b is the signal-to-noise ratio and $\lfloor x \rfloor$ is the function that converts the value of x into the largest integer less or equal to x.

The average bit error probability considering a Nakagami-m fading channel is evaluated through the integration [14]:

$$P_{BER} = \int_0^8 P_{bn} \times P_\gamma(\gamma) d\gamma, \tag{6}$$

where P_γ is the PDF of the SNR in the receiver, and P_{bn} is the average bit error probability for rectangular M-QAM modulations, given by Eq. (3) for an AWGN (Additive White Gaussian Noise) channel.

4 Throughput Calculation

To compute the normalized throughput of a hybrid system, the following factors were considered: the ratio between the number of bits per symbol of the current modulation and the number of bits per symbol of a reference modulation; the fraction of information bits transmitted in the link, expressed by the ratio between the number of information bits in a packet, k, and the total number of bits in a packet, n; and the probability of a packet contains no errors or contain only correctable errors. Thus, the throughput in a hybrid system is defined by [5]:

$$\eta = \frac{\log_2 M_i}{\log_2 M_r} \times \frac{k}{n} \times (1 - PER), \tag{7}$$

where M_i is the order (number of symbols in the constellation) of the current modulation and M_r is the order of a reference modulation.

5 Power Metric Calculation

The power metric was proposed by Giessler et al. in [16]. This performance metric is defined as the ratio between the throughput and the delay in the system [16]. In [17], the authors state that this parameter is effective in measuring the impact of the delay in the system performance, which allows greater efficiency in the configuration of transmission parameters. The power metric is computed by Expression (8):

$$P_M = \frac{\eta}{E(T)}, \tag{8}$$

where η is the normalized throughput, defined by Expression (7), and $E(T)$ is the mean time required to transmit a data message correctly. The calculation of the delay in the system is based on the methodology proposed in [8], which in turn was based on the approach presented in [15]. In this methodology, the following was considered: the system uses TDMA (Time Division Multiple Acess) with a frame composed by X slots; the data message is transmitted in Z packets; an additional error control scheme in the data message level, based on ARQ (Automatic Repeat Request) is used, and the message is retransmitted if received with error. Thus, the delay is computed by [5, 8]:

$$E(T) = \frac{n_s[(Z-1)X+1]}{P_{dn}\beta_n B} + \frac{Kn_s X}{\beta_n B}\frac{(1-P_{dn})}{P_{dn}}. \tag{9}$$

where X is the number of time slots in a frame, K is the average number of frames between the end of an incorrect transmission and the start of the retransmission of a data message, β_n is the bandwidth efficiency of the current modulation, given by $\beta_n = \log_2 M$ [bps/Hz], B is the bandwidth of the channel, and P_{dn} is the probability of receiving a correct data message, given by [15]:

$$P_{dn} = (1 - PER)^Z \tag{10}$$

6 Numerical Results

In this section, we present the numerical results for the computation of the optimal switching points, considering as performance criteria throughput and power metric. The analysis was performed as a function of the E_S/N_0 ratio (symbol energy to noise power density ratio). Following [11], the number of information bits in a packet was defined as $k = 424$, and the reference modulation used was 256-QAM. The influence of the channel model was obtained by varying the m parameter of the Nakagami, placing $m = 0.5, 1, 2, 3,$ and 10 [11]. We used (7) to calculate the throughput and (8) for power metric, all computations were made using Mathcad®.

6.1 Optimum Switching Points Based on the Throughput Criterion

In this subsection, the performance of the hybrid technique proposed in [5] is analyzed under a Nakagami-m fading channel. This hybrid technique considers that an adaptive modulation is associated with adaptive FEC. The analysis in this section is based on the throughput criterion, obtained for the modulations 256, 128, 64, 16, and 8-QAM, when an adaptive FEC code is added to keep the PER below 0.1 (fixed threshold). In other words, for each modulation, we defined a BCH code as a function of E_S/N_0 to keep the PER equal to or lower than 0.1, and then we compute the throughput using (7).

Figure 1 show the throughput curves for $m = 0.5, 1, 3,$ and 10, respectively. The optimal switching points are defined by crossing the throughput curves between neighboring modulations and are shown in Table 1. It has been noticed that some switching points between neighboring modulations are close to each other. We can verify, for example, that for the neighboring 256 and 128-QAM modulations, and for the neighboring 64 and 32-QAM modulations, the switching points assume very close

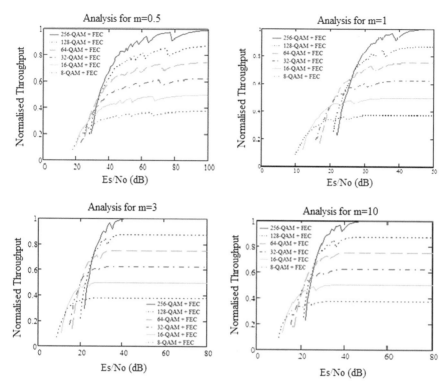

Fig. 1 Normalized throughput as a function of E_S/N_0 for the hybrid adaptive scheme, $m = 0.5$, 1, 3 and 10, $M = 8$, 16, 32, 64, 128 and 256 QAM

values. The switching point between 128 and 64-QAM occurs before the switching point between 256 and 128-QAM. Also, the switching point between modulations 32 and 16-QAM occurs before the switching point between 64 and 32-QAM. So, considering the throughput as the performance criterion, 128 and 32-QAM modulations can be discarded.

Based on the results presented in Table 1, one can conclude that the optimum switching points depend on the behavior of the communication channel, modeled by the m parameter of the Nakagami-m distribution. Thus, the definition of the switching points must consider the model of the channel.

6.2 Optimum Switching Points Based on the Power Metric Criterion

In this subsection, we analyzed the influence of the channel model in the determination of the ideal switching points considering the power metric criterion. To compute

Table 1 Optimum switching points (hybrid adaptive FEC and modulation)

Switching	m	Optimum switching E_S/N_0 (dB)
To 256 for 128-QAM	0.5	32
	1	27
	3	26
	10	25
To 128 for 64-QAM	0.5	34
	1	28
	3	25
	10	24
To 64 for 32-QAM	0.5	26
	1	21
	3	19
	10	19
To 32 for 16-QAM	0.5	27
	1	22
	3	19
	10	18
To 16 for 8-QAM	0.5	21
	1	15
	3	13
	10	12

the delay we used (9) setting $Z = 1$, $X = 10$, $B = 1$ MHz, and $K = 1$. Again, for each modulation, first we defined a BCH code as a function of E_S/N_0 to keep the PER equal to or lower than 0.1, and then we compute the throughput using (7), the delay using (9) and the power metric using (8).

Figure 2 show the power metric curves as a function of the E_s/N_0, and the Nakagami parameter varying in $m = 0.5$, 1, 3, and 10, respectively. Again, we considered the modulations 256, 128, 64, 16, and 8-QAM. The crossing point of the corresponding power metric curves determines the ideal switching points. Table 2 summarizes the different crossing points obtained.

Based on the results presented in Table 2, similar to what happened in the previous criterion, we can see that the optimum switching points vary with the order of diversity m. Also, comparing the results presented in Tables 1 and 2, we can conclude that the optimum switching points depend on the performance criterion.

It is also possible to verify that there is some similarity with the behavior observed to the throughput criterion. For example, for $m = 0.5$, the switching point between modulations 128 to 64-QAM occurs before the switching point between modulations 256 to 128-QAM. The same is true for the switching point between 32 to 16-QAM modulations in relation to the switching between 64 to 32-QAM modulations. Also,

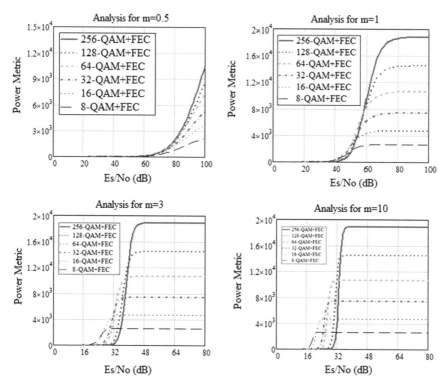

Fig. 2 Power metric as a function of E_S/N_0, m = 0.5, 1, 3 and 10, M = 8, 16, 32, 64, 128 and 256 QAM

for $m = 1$, the switching point between modulations 128 to 64-QAM occurs before the point between modulations 256 to 128-QAM, and the same happens for the point between 32 to 16-QAM in relation to 64 to 32-QAM. Thus, it can be concluded that, when m assumes the values m = 0.5 or 1, modulations 128 and 32-QAM should not be used in the hybrid adaptive FEC modulation system when the power metric criterion is considered.

7 Conclusion

The analysis of switching points is a key factor in maximizing the performance of adaptive modulation or FEC systems. In this paper, we computed the ideal switching points of an adaptive hybrid technique that combines adaptive modulation and adaptive FEC techniques, considering a Nakagami-m fading channel. By varying the m parameter, we investigated the influence of the channel model in the definition of the optimum switching points. We computed the optimum switching points considering two different performance criteria: throughput and power metric.

Switching	m	Optimum switching E_S/N_0 (dB)
Table 2 Optimum switching point and power metric		
To 256 for 128-QAM	0.5	86
	1	55
	3	39
	10	33
To 128 for 64-QAM	0.5	98
	1	58
	3	37
	10	31
To 64 for 32-QAM	0.5	47
	1	45
	3	33
	10	27
To 32 for 16-QAM	0.5	66
	1	50
	3	31
	10	25
To 16 for 8-QAM	0.5	46
	1	34
	3	26
	10	21

Based on the results, the paper concluded that the ideal switching points depend on: the performance criterion defined to the system (throughput or power metric); the channel model, defined by the fading factor (m); the current modulation. Furthermore, it was also possible to observe that some switching points between neighboring modulations are close to each other, indicating that some modulations can be omitted when implementing a hybrid adaptive system; that is the case of modulations 128-QAM and 32-QAM.

References

1. Bandiri MSY, Braga R, Spadoti DH (2017) Analytical comparison of the performance of adaptive modulation and coding in wireless network under Rayleigh fading. J Microw Optoelectron Electromagn Appl 16(3):723–735
2. Rhee D, Kwon JH, Hwang HK, Kim KS (2006) Adaptive modulation and coding on multipath Rayleigh fading channels based on channel prediction. In: 8th international conference advanced communication technology—ICACT. IEEE, Phoenix Park, South Korea, pp 195–199

3. Baguda Y, Fisal N, Syed S, Latiff L, Yusof S, Rashid R, Sani D (2010) Adaptive FEC error control scheme for wireless video transmission. In: 12th international conference on advanced communications technology (ICACT). IEEE, Phoenix Park, South Korea, pp 565–569
4. Lin S, Costello DJ (1983) Error control coding: fundamentals and applications. Prentice Hall
5. Brito JMC, Bonatti IS (2002) An analytical comparison among adaptive modulation, adaptive FEC, adaptive ARQ and hybrid systems for wireless ATM networks. In: The 5th international symposium on wireless personal multimedia communications, vol 3. IEEE, Honolulu, HI, USA, pp 1034–1038
6. Brito JMC, Bonatti IS (2002) Analysing the switching points in wireless ATM networks that use adaptive FEC schemes. In: Proceedings the second international symposium on communications and information technology (ISCTIT02), Pattaya, Thailand, pp 305–308
7. Bandiri MSY, Brito JMC (2015) Analyzing the Optimum switching points for adaptive FEC in wireless networks with Rayleigh fading. In: Proceedings of the ICN 2015, the fourteenth international conference on networks, Barcelona, Spain, pp 23–28
8. da Silva APTR, Brito JMC (2016) Analysis of the optimum switching points in an adaptive modulation system in a Nakagami-m fading channel considering throughput and delay criteria. Int J Adv Telecommun 9(3 & 4):44–54
9. Foukalas F, Khattab T, Poor HV (2013) Adaptive modulation in multi-user cognitive radio networks over fading channels. In: 8th international conference on cognitive radio oriented wireless networks (CROWNCOM). IEEE, Washington, DC, USA, pp 226–230
10. Bandiri MSY, Brito JMC (2014) Analyzing the optimum switching points for adaptive modulation in wireless networks with Rayleigh fading. In: 6th IEEE Latin American conference on communications. IEEE, Cartagena, Colombia, pp 1–6
11. da Silva APTR, Brito JMC (2017) Analysis of adaptive modulation performance in networks with multiple access Slotted Aloha. In: The 31st international conference on information networking (ICOIN 2017). IEEE, Da Nang, Vietnam, pp 210–215
12. Wang Z, Giannakis GB (2003) A simple and general parameterization quantifying performance in fading channels. IEEE Trans Commun 51(8):1389–1398
13. Nakagami M (1960) The m-distribution—a general formula of intensity distribution of rapid fading. In: Proceedings of a symposium held at the University of California. Elsevier, Los Angeles, CA, USA, pp 3–36
14. Cho K, Yoon D, Jeong W, Kavehrad M (2001) BER Analysis of arbitrary rectangular QAM. In: Conference record of thirty-fifth asilomar conference on signals, systems and computers (Cat. No. 01CH37256), vol. 2. IEEE, Pacific Grove, CA, USA, pp 1056–1059
15. Cho K, Yoon D (2002) On the general BER expression of one- and two-dimensional amplitude modulations. IEEE Trans Commun 50(7):1074–1080
16. Giessler A, Hänle J, König A, Pade E (1978) Free buffer allocation—an investigation by simulation. Comput Netw (1976) 2(3):191–208
17. Rubem TB, Cardieri P (2013) Métrica de Eficiência em Redes de Rádio Cognitivo com Fila e Retransmissão Garantida. In: XXI SBrT, Fortaleza, CE, Brazil, pp 1–5

Performance Evaluation of Linear LoRa Network Protocol

Ngoc Huy Nguyen, Dinh Loc Mai, and Myung Kyun Kim

Abstract The Internet of Thing (IoT) applications such as industrial, agriculture, smart home applications are demanding strong requirements about distances and power consumption. LoRa (Long Range) is a novel solution for these requirements. LoRa is a spread spectrum modulation technique derived from chirp spread spectrum (CSS) technology. LoRa technology can be applied to public, private or hybrid networks. Multi-hop LoRa linear protocol is a new protocol using LoRa technology. In the paper, we evaluate the performance of this protocol by data reception rate, successful tree construction rate in different condition. Our results show that LoRa linear protocol has high reliability and easy to apply to the real-environment.

Keywords LoRa network · IoT · Multi-hop LoRa · Low-power

1 Introduction

The Internet of Thing (IoT) is revolutionizing our life and work with the smart application which applied in the home, industrial and agriculture. Wireless sensor network (WSN) is often a technology used within an IoT system. WSN is a network for gathering sensor data from sensor node then storing in database storage. There some technologies are applied for wireless sensor network such as Zigbee, Wi-Fi, Bluetooth, Cellular, LoRa [1–3] showed in Fig. 1.

In our work, we focus is on LoRa technologies. LoRa is new technology for long-distance and low-power consumption and the frequency bands used for sub-GHz. LoRa technology has some important parameter which value as shown in Table 1.

- Spreading factor (SF): is the ratio between the symbol rate and chip rate. Both transmitter and receiver must be known spreading factor value otherwise are orthogonal to each other.

N. H. Nguyen · D. L. Mai · M. K. Kim (✉)
University of Ulsan, Daehak-Ro 93, Nam-Gu, Ulsan 44610, Korea
e-mail: mkkim@ulsan.ac.kr

Fig. 1 Wireless
technologies

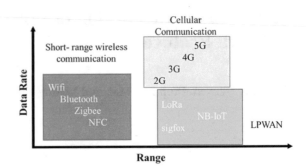

Table 1 Parameter value

Parameters	Values
Spreading Factor (SF)	6 to 12
Coding Rate (CR)	4/5, 4/6, 4/7, 4/8
Channel Bandwidth (BW)	125, 250, 500 kHz

- Bandwidth (BW): is the range of frequencies in the transmission band. Higher BW reduces sensitivity at transceiver of receiver but it increase data rate. In LoRa transceiver (SX1272) has three available BW such as 125, 250 and 500 kHz.
- Coding Rate (CR): is Forward Error Correction (FEC) rate. It used by LoRa modulation to recovery error bit of information when packet is transmitted between transmitter and receiver.

In LoRa technology, it has some advantages:

- Transmission range is much longer than other technologies, maximum range of transmission is longer than 10 km and increases spreading factor the transmission range increases.
- Sensitivity in LoRa is very much lower than other technologies as shown in Table 2, example in Wi-Fi, senility is −90dBm.
- LoRa technology applied to challenging environment as in tunnels [4] and top of tower [5], but in this case, another technologies such as Wi-Fi, Bluetooth are very difficult to apply.
- LoRa technique is not effected by multi-path fading [6].

Table 2 Semtech SX1272 LoRa receiver sensitivity (dBm)

SF BW	6	7	8	7	10	11	12
125 kHz	−121	−124	−127	−130	−133	−135	−137
250 kHz	−118	−122	−125	−128	−130	−132	−135
500 kHz	−111	−116	−119	−122	−125	−128	−129

Table 3 LoRa data rate (bps)

SF BW	7	8	9	10	11	12
125 kHz	5468	3125	1757	976	537	292
250 kHz	10,937	6250	3515	1953	1074	585
500 kHz	21,875	12,500	7031	3906	2148	1171

But LoRa has disadvantage is lower data rate than other technologies as shown in Table 3 with CR = 4/5 and BW = 125 kHz. In Zigbee it is 250kps, Bluetooth can reach 3 Mbit/s.

Some researcher makes LoRa protocol based on LoRa technology. Only LoRaWAN protocol is open-source protocol, it developed by IBM with name is the IBM LoRaWAN C-library (LMiC). Many researcher make experiment with LoRaWAN protocol to evaluate performance of this protocol [7–9]. Thus, our work evaluates performance of our protocol: LoRa linear protocol [10]. In our protocol, it high reliability protocol and easily apply to challenging environment.

The remainder of this paper is organized as follows. Section 2: related work. In Sect. 3, describe our experiments, we evaluate the experimental performance of the proposed protocol in terms of network construction probability, packet reception reliability. Finally, we conclude the paper in Sect. 4.

2 Related Work

In Lora technology, we have some parameters effect to performance of LoRa communication such as transmission power, SF. When we increase transmission power and SF value, transmission range is increases. In our experiment, we focus SF7 because it is suitable for our experiment as distance between nodes is 0–1 km and with SF7, we have highest data rate when comparing with other spreading factor. First experiment, we evaluate successful network construction capability. Second experiment, we evaluate packet reception rate when we deployed node in building and different distances. With power transmission, we choose 0 dBm, because it is low power transmission. In sensor wireless sensor network, especially TSCH network, devices choose this value to the default value for power transmission and we can extend life-time for LoRa node. Other researcher choose 13, 20 dBm [11] for power transmission value, it is too much and not suitable for low-power network. Our experiment focus when we apply the protocol to control light system or monitor temperature value from different floor in the building.

2.1   LoRa Linear Network Protocol

In Lora linear network protocol, it consist of the network construction period (NCP), the upward transmission period (UTP), and the downward transmission period (DTP) as shown in Fig. 2. Every node have unique ID, ID of node increases one from zero, in which node has ID equal zero is sink node. In UTP and DTP, the length of slot-frame equals number of deployed node.

- In NCP, sink node first transmits INIT(S_{ID},Timeslot,N) packet where S_{ID} is the ID of the sensor node, Timeslot is current timeslot when node transmits INIT packet, and N is number of timeslot in the NCP. Other nodes try to listen INIT packet. After receiving INIT packet, if S_{ID} is the previous ID of receiver node, receiver node synchronizes time with sender node and transmits INIT packet in next timeslot otherwise, receiver node tries to listen other INIT packet. The sender node tries to listen INIT packet at next timeslot to guarantee that next ID node received INIT packet from this node. If the sender node doesn't receive INIT packet, the sender node will transmit INIT packet one more time to increase successful network construction capability. This process repeats until leaf-node receives INIT packet. After network construction period finish, the protocol starts data transmission period.
- In UTP period, leaf node transmits data packet to previous hop node, after receiving data packet, receiver node combines its own data with received data then transmit packet to previous hop node of receiver. This process repeats until receiver node is sink node as shown in Fig. 3.

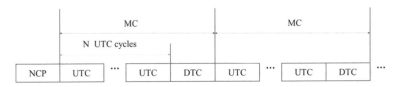

Fig. 2 LoRa Linear network structure

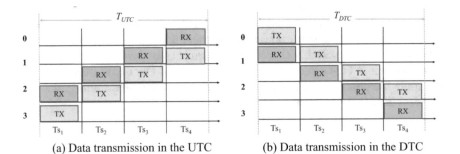

(a) Data transmission in the UTC   (b) Data transmission in the DTC

Fig. 3 Data transmission period

- After n UTC, it starts DTP, sink node transmits command packet to next-hop node, receiver node forward this packet until receiver node as leaf node. In UTP period, all of node are resynchronized time together.

In LoRa linear protocol, the node chooses the previous ID node is the parent node. Based on number deployed node and ID of node, every node determine timeslot node transmits, receives data packet or command packet, with example with 4 nodes as shown in Fig. 3. Otherwise, node is in idle state.

3 Experiment Method and Results

3.1 Hardware Setup

Every sensor node contains GPS sensor, micro-SD card and Multi-tech mDot module as shown in Fig. 4b. We developed LoRa linear network with Multi-tech mDot module, it consist of Semtech SX1272 LoRa transceiver that work in the 920 kHz and an STM32F411RET processor. Devices is supplied by 3.7 V Li-on battery with capacity of 3.5Ah that can be charged via charger module. Sink node connects directly to PC, network sets up as shown in Fig. 4a. To evaluate the performance of the proposed protocol, we conducted an experiment by deploying 4 nodes with the set of parameters in Table 4.

(a) LoRa linear network   (b) LoRa node prototype

Fig. 4 Hardware setup

Table 4 Experiment configuration parameters

Number of deployed nodes	4	Coding Rate (CR)	CR 4/5
Band Width (BW)	125 kHz	Preamble length	8 symbols
Transmission power	0 dBm	Header mode	explicit
Spreading factor	SF7	CRC	enable

3.2 Experimental Setup

Every node reads configuration from micro-SD card. In micro-SD card, we configure ID of node for every node. In SD card from sink node, we configure number of deployed node.

SF can be set from 6 to 12, in experiment we choose SF7 because range of transmission if we use SF7 is 0–1 km, it is most suitable for distance between of nodes.

We make experiment in university. To evaluate radio successful tree construction, we make experiment in Scenario 1. To get results, we launch protocol 220 times then get number of successful tree construction.

- Scenario 1: all nodes are deployed outside.

Indoor environments are more challenging for wireless communications including LoRa communication. To evaluate correctly protocol performance, we make all of experiments for packet-reliability are indoor experiments, sink node was deployed in one room. The distance from sink node to door is 3 m. Each UTC, each node except sink node transmits one, in DTC each node except leaf-node transmits one packet. To get results data packet reception rate we consider data packet reception rate which sink node received from leaf-node. With command packet reception rate, we consider command packet what leaf-node received from sink node. In scenario 2, 3, 4, 5 we make three times for each scenario. In one test, leaf-node transmits data packet 120 times and sink node transmits command packet 120 times.

- Scenario 2: nodes are deployed in one lab room.

- Scenario 3, 4, 5: sensor nodes are deployed at the lobby on the different floors of the building as shown in Fig. 5.

In Scenario 1, after 220 times test, we get 214 times network construction is successful. It is about 97.3% successful. Successful rate is very high (Fig. 6).

In Fig. 7, command packet reception rsate in different environments is very high, it is greater than 98.3%. Because in DTC, every node choose each timeslot

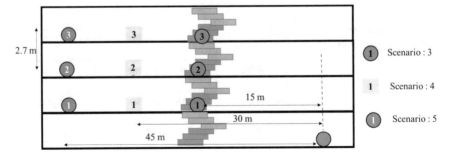

Fig. 5 Deployed node in Scenario 3, 4, 5

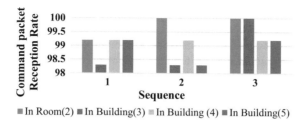

Fig. 6 Command packet reception rate in Scenario 2, 3, 4, 5

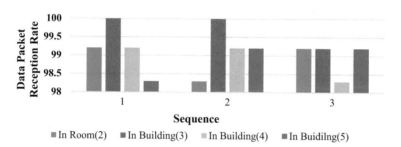

Fig. 7 Data packet reception rate in Scenario 2, 3, 4, 5

to transmit command packet and no-collision occurs. The command packet is from sink then forward to other node until leaf-node. Thus, we can use protocol for control application in same environment with experiment environment.

In Fig. 7, data packet reception rate in different environments is very high, it is greater than 98.3%. Because in UTC, every node chooses each timeslot to transmit command packet and no-collision occurs. From the obtained results, we can conclude that LoRa linear network protocol is high reliability protocol and we can apply the protocol to high reliability application.

4   Conclusion

LoRa technology is a long-range and low-power technology apply to the "Internet of Things". We developed LoRa linear network protocol with Multi-tech mDot module in Mbed OS. The purpose of this paper is to analyze the performance of the LoRa linear protocol. In LoRa linear protocol, every sensor node transmits data, command packet at different timeslot so no-collision occurs. We first evaluate percentage of successful network construction as shown the protocol easily apply to real-environment. Second, we evaluate reliability of the protocol in different environment. In result, packet reception rate is greater than 98.3% so the protocol is high reliability. Based on experiment, with distance from 0 to 1 km, in LoRa technology, we should use SF7 for transmitting data packet. The protocol is suitable for low data-rate and high reliability application.

As future work, we plan to evaluate LoRa linear protocol about energy consumption in other environment such as in tunnel, basement, long-distance outside.

References

1. Raza U, Kulkarni P, Sooriyabandara M (2017) Low power wide area networks: an overview. IEEE Commun Surv Tutor 19(2):855–873
2. Sharma H, Sharma S (2014) A review of sensor networks: technologies and applications. In: 2014 recent advances in engineering and computational sciences (RAECS), pp 1–4
3. Mekki K, Bajic E, Chaxel F, Meyer F (2019) A comparative study of LPWAN technologies for large-scale IoT deployment. ICT Express 5(1):1–7
4. Abrardo A, Pozzebon A (2019) A multi-hop LoRa linear sensor network for the monitoring of underground environments: the case of the Medieval Aqueducts in Siena, Italy. Sensors 19(2):402
5. Petajajarvi J, Mikhaylov K, Roivainen A, Hanninen T, Pettissalo M (2015) On the coverage of LPWANs: range evaluation and channel attenuation model for LoRa technology. In: 2015 14th international conference on ITS Telecommunications (ITST), pp 55–59
6. Nguyen TT, Oh H (2018) A receiver for resource-constrained wireless sensor devices to remove the effect of multipath fading. IEEE Trans Indus Electron 65(7):6009–6016
7. Bouguera T, Diouris J-F, Chaillout J-J, Jaouadi R, Andrieux G (2018) Energy consumption model for sensor nodes based on LoRa and LoRaWAN. Sensors 18(7):2104
8. Lavric A, Popa V (2018) Performance evaluation of LoRaWAN communication scalability in large-scale wireless sensor networks. Wirel Commun Mob Comput, vol 2018
9. Darbandi A, Kim MK (2019) Path collision-aware real-time link scheduling for TSCH wireless networks. KSII Trans Internet Inf Syst 13(9):4429–4445
10. Cong Tan D, Kim M-K (2018) Reliable multi-hop linear network based on LoRa. Int J Control Autom 11:143–154
11. Cattani M, Boano CA, Römer K (2017) An experimental evaluation of the reliability of lora long-range low-power wireless communication. J Sens Actuator Netw 6(2):7

Extending the Performance Analysis of SHARP—Spectrum Harvesting with ARQ Retransmission and Probing in Cognitive Radio

Nicolas Wilson Ribeiro Rocha◉ and Jose Marcos Camara Brito◉

Abstract This paper proposes an extension of the method entitled Spectrum Harvesting with ARQ Retransmission and Probing (SHARP), an underlay cognitive radio network (CRN) technique where the secondary pair listens to the primary Automatic Repeat Request (ARQ) feedback to glean information about the primary channel and probe it getting additional information of the relative cross channel strength. This probing mechanism and two varieties of spectrum sharing, named conservative and aggressive SHARP, are used depending on the desired secondary user interference in the primary. A completely new modeling was made considering more retransmissions and, therefore, analyzing new probabilities and regions of operation. The influence in the performance of the scheme by increasing the number of retransmissions is studied based on numerical results and some insight about the evaluation of a generic scenario prediction are provided, as long as a trade-off conclusion about the viability of increasing or not the number of retransmissions in such schemes.

Keywords ARQ · Cognitive radio · Outage probability · Spectrum sharing · SHARP · CRN

1 Introduction

Cognitive radio techniques have become an important solution for solving spectral congestion problems. The literature involving the coexistence of primary and secondary users grows with the interest of increasing spectrum efficiency [1].

A fundamental requirement of a cognitive radio is to be aware of the primary users (PU) in the environment, inflicting minimal interference on them. The theoretical and

N. W. R. Rocha · J. M. C. Brito (✉)
INATEL – National Institute of Telecommunications, Santa Rita do Sapucai, MG 37540000, Brazil
e-mail: brito@inatel.br

N. W. R. Rocha
e-mail: nicolasrocha@get.inatel.br

© The Author(s), under exclusive license to Springer Nature Singapore Pte Ltd. 2021
H. Kim et al. (eds.), *IT Convergence and Security*, Lecture Notes
in Electrical Engineering 712, https://doi.org/10.1007/978-981-15-9354-3_8

practical problems involved in cognitive radio have proved to be very challenging. Zhao and Sadler [2] summarizes some results in dynamic spectrum sharing.

Several studies present possible improvements in CR techniques and its viability. Chandwani, Jain, and Vyava [3] discuss the primary user signal and channel noise conditions in a throughput comparison. Park, Kim, Lim, and Song [4] explore CR techniques by searching for new relevant channels for communication. Barnawi [5] focused on a novel approach by sensing the radio environment using a wideband chirp signal. Hattab [6] shows the advances in multiband spectrum sensing techniques for cognitive radios networks (MB-CRNs).

Furthermore, ARQ retransmission protocols have being exploited in cognitive radio. Li, Zhang, Nosratinia, and Yuan introduce the SHARP scheme [7], a method of underlay cognitive radio where the secondary users (SU) listen to the primary ARQ feedback to glean information about the primary channel. More recently, Bahgat presents a design of a low cost cognitive radio platform for demonstration and testing purposes [8], useful for validating the theory study. This paper aims to extend the performance analysis of the SHARP scheme previously presented in [7]. In this extension, three retransmissions are considered instead of two retransmissions as in [7] by doing a completely new modeling. The goals are to analyze the influence of the number of retransmissions in the performance of the scheme and have some insight about the evaluation of a generic scenario prediction.

The remainder of this paper is organized as follows. Section 2 introduces the system model. In Sects. 3 and 4, the operating regions are presented, along with their corresponding probabilities, and a flow chart summarizes the developed algorithm for operation region discovery. Furthermore, the throughput analysis of the proposed schemes is given in Sect. 5. Finally, mathematical results and conclusion remarks are shown in Sects. 6 and 7, respectively.

2   System Model

The system model used is illustrated in Fig. 1 [7]. The primary transmitter occupies the channel continuously; therefore, the secondary transmitter can use the channel only through spectrum sharing. The channel gains are g_{ij} from a transmitter i to

Fig. 1 ARQ-based spectrum sharing transmission system

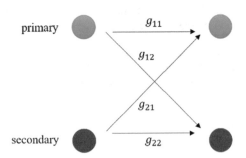

a receiver j, where the subscript value 1 denotes the primary and 2 denotes the secondary. Channel gains obey an exponential distribution with mean λ.

A slow fading scenario is considered with the channel gain assumed approximately constant over several transmission intervals, but subject to change over much larger time scales. The primary transmitter operates at a constant power P_p and the nominal spectral efficiency of R_p bits/s/Hz. If the first transmission at this rate and power is not successful (indicated by a NACK from the primary receiver), the same packet is re-transmitted at the same power. The receiver combines the two packets and, if there is a failure in decoding after two transmissions, the primary may try a third time before declaring outage and moving to the next package. Therefore, the analysis is limited to three ARQ rounds.

The secondary transmitter has a power constraint of P_s and a nominal spectral efficiency of R_s bits/sec/Hz. Its receiver does not generate ARQ feedback. This intermittent transmission is characterized by the throughput related to its outage.

The system model is a simplified one as in [7] to focus on the most important aspects of the ideas and for analysis simplicity. The contribution of this paper is to extend the analysis presented in [7] in order to verify the influence of the maximum number of retransmissions in the ARQ scheme in the performance of the system. Thus, in this paper there is a maximum of three retransmissions, instead of two, as in [7]. To analyze this new scenario, new modeling was necessary including the definition of new operation regions along with their respective probabilities and new throughput computation.

3 Operating Regions

The premise of the SHARP method is allowing a secondary transmitter to share the primary channel without needing much channel-state information. It uses the ARQ from primary receiver to exploit opportunities for transmissions whenever possible. The resulting interference may not drive the primary into outage.

The channel states are not directly available to the secondary due to practical reasons. It can only observe the ACK/NACK from the primary receiver. Since it also knows its own transmissions, it is possible to know whether the ACK/NACK from the primary was under secondary interference or not.

Secondary transmissions are allowed depending on which region the system is operating, defined by the relative strength of the direct channel gain, g_{11}, and the cross channel gain, g_{21}. From the Shannon Capacity Theorem, the signal-to-noise ratio in the primary and secondary links are, respectively, $\gamma_p \triangleq 2^{R_p} - 1$ and $\gamma_s \triangleq 2^{R_s} - 1$. These parameters are the thresholds used to characterize the operating regions. There are ten regions instead of only six as in [7] because more retransmissions are considered.

In telecommunications, the maximum-ratio combining (MRC) is a commonly used method of diversity combining. The signals of each channel are added and the fading considered being independent in order to increase the SNR. The result is a signal which SNR is the sum of the SNR at each element [9].

To define the operating regions probabilities, MRC is used. The ten possible operating regions, named $S_1 - S_{10}$, are:

1. The primary channel supports the rate in one transmission despite secondary interference. It is the best possible scenario. Under this condition, the secondary may always transmit.

$$\frac{P_p g_{11}}{P_s g_{21} + N} > \gamma_p \tag{1}$$

2. The primary channel can support its rate in one transmission without interference, but needs two transmissions to succeed in the presence of interference. Under this condition, the secondary can transmit all time without driving the primary into outage, but may decrease its throughput.

$$\frac{\gamma_p}{2} < \frac{P_p g_{11}}{P_s g_{21} + N} < \gamma_p \tag{2}$$

$$\frac{P_p g_{11}}{N} > \gamma_p \tag{3}$$

3. The primary channel can support its rate in one transmission if there is no interference, but needs three transmissions to succeed in the presence of interference. Under this condition, the secondary can transmit all time without driving the primary into outage, but also may decrease its throughput.

$$\frac{\gamma_p}{3} < \frac{P_p g_{11}}{P_s g_{21} + N} < \frac{\gamma_p}{2} \tag{4}$$

$$\frac{P_p g_{11}}{N} > \gamma_p \tag{5}$$

4. The primary channel again support its rate in one transmission if there is no interference, but it cannot succeed even with three transmissions. Under this condition, the secondary can transmit all time without driving the primary into outage, but it will continue to decrease its throughput.

$$\frac{P_p g_{11}}{P_s g_{21} + N} < \frac{\gamma_p}{3} \tag{6}$$

$$\frac{P_p g_{11}}{N} > \gamma_p \tag{7}$$

5. The primary channel can support its rate in two free interference transmissions; it can also succeed in two transmissions if only one of them is subject to

interference. Under this condition, the secondary should transmit on only one transmission interval without any effect on the primary.

$$\frac{P_p g_{11}}{N} + \frac{P_p g_{11}}{P_s g_{21} + N} > \gamma_p \tag{8}$$

$$\frac{\gamma_p}{2} < \frac{P_p g_{11}}{N} < \gamma_p \tag{9}$$

6. The primary channel can support its rate in two free interference transmissions; it can also succeed in three transmissions as long as only two of them is subject to interference. Under this condition, the secondary should transmit on two transmissions intervals without driving the primary into outage, but it will continue to decrease its throughput.

$$\frac{P_p g_{11}}{N} + \frac{2 P_p g_{11}}{P_s g_{21} + N} > \gamma_p \tag{10}$$

$$\frac{\gamma_p}{2} < \frac{P_p g_{11}}{N} < \gamma_p \tag{11}$$

7. The primary channel can support its rate in two free interference transmissions; it cannot support its rate with any interference. Under this condition, the secondary may remain silent.

$$\frac{P_p g_{11}}{N} + \frac{2 P_p g_{11}}{P_s g_{21} + N} < \gamma_p \tag{12}$$

$$\frac{\gamma_p}{2} < \frac{P_p g_{11}}{N} < \gamma_p \tag{13}$$

8. The primary channel can support its rate in three free interference transmissions; it can also succeed in three transmissions as long as only one of them is subject to interference. Under this condition, the secondary should transmit on only one transmission interval without any effect on the primary.

$$\frac{2 P_p g_{11}}{N} + \frac{P_p g_{11}}{P_s g_{21} + N} > \gamma_p \tag{14}$$

$$\frac{\gamma_p}{3} < \frac{P_p g_{11}}{N} < \frac{\gamma_p}{2} \tag{15}$$

9. The primary channel can support its rate in three free interference transmissions; it cannot support its rate with any interference. Under this condition, the secondary may remain silent.

$$\frac{2P_p g_{11}}{N} + \frac{P_p g_{11}}{P_s g_{21} + N} < \gamma_p \tag{16}$$

$$\frac{\gamma_p}{3} < \frac{P_p g_{11}}{N} < \frac{\gamma_p}{2} \tag{17}$$

10. The primary channel gain is so small that forces the primary into outage even with retransmission and no interference. Under this condition, the secondary should transmit at every time.

$$\frac{P_p g_{11}}{N} < \frac{\gamma_p}{3} \tag{18}$$

4 Probing and Discovery Mechanism

In some conditions, the SU may remain silent to avoid pushing the PU into outage or it can transmit without causing any outage, but decreasing the effective PU throughput due to retransmissions. In the *aggressive* SHARP, the secondary may transmit whenever possible, whereas in the *conservative* SHARP, it can only transmit when there is no effect on the primary. The probing and discovery mechanism are characterized by secondary transmission decisions. The transmission modes are as follows:

$T_0 = \{$primary transmits new packet secondary keeps silent$\}$
$T_1 = \{$primary repeats old packet secondary keeps silent$\}$
$T_2 = \{$primary transmits new packet secondary transmits$\}$
$T_3 = \{$primary repeats old packet secondary transmits$\}$

Using the above notation, the mechanism is systematically implemented for aggressive and conservative SHARP as illustrated in Figs. 2 and 3, respectively. Starting from the root of the tree, the secondary stays silent for the first transmission and observes the primary ACK/NACK. Each detection traces a path from the top

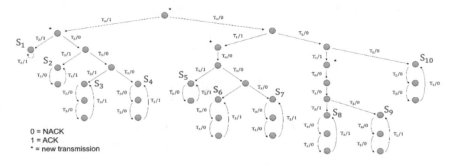

Fig. 2 Flow chart for the *aggressive* SHARP

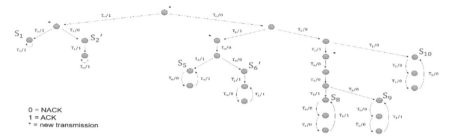

Fig. 3 Flow chart for the *conservative* SHARP. The regions S_2, S_3, S_4 from aggressive SHARP which resulted in throughput degradation are now modified into a single one named S_2', as well as regions S_6 and S_7 which united into S_6'. In these, no secondary transmission is allowed

of the tree to one of the ten possible leaves. When a region is found, a transmission loop starts and continues until the channel conditions change. Occasionally, due to channel fading, the secondary may redo the probing and restart the process. It is important to note that neither aggressive nor conservative SHARP produce any primary outage. The main difference resides in the fact that aggressive SHARP may occasionally slow down the primary by forcing it in some occasions to retransmit, causing some degradation compared to conservative SHARP.

5 Throughput Analysis

Some analytical results for the cognitive radio system are presented in this section when the *aggressive* and *conservative* SHARP scheme is applied. The probabilities of all ten regions are calculated based on the threshold limits. Outage probability for the primary and secondary links are also studied. Finally, the achievable throughput for both users is derived.

5.1 Operating Region Probabilities

For simplicity, the same premises assumed in [7] are used: (I) Gaussian noise variance N equals to 1; (II) the channel gain is decoupled by the mean value λ and random variable x, i.e. $g_{11} \triangleq \lambda_{11}x_{11}$ and $g_{21} \triangleq \lambda_{21}x_{21}$, where x_{11}, x_{21} follow exponential distribution with unit mean. These decompositions are very common in wireless communications for analysis convenience in order to isolate and normalize the effect of small-scale fading.

The probability of the channel gains residing in each of the ten regions are as follows:

$$P\{S_1\} = P\left\{\frac{P_p g_{11}}{P_s g_{21} + 1} > \gamma_p\right\} = \int_0^\infty \left[e^{-\gamma_p \frac{(1+P_s \lambda_{21} y)}{\lambda_{11} P_p}} e^{-y}\right] dy \qquad (19)$$

$$P\{S_2'\} = P\left\{P_p g_{11} \geq \gamma_p, \frac{P_p g_{11}}{P_s g_{21} + 1} < \gamma_p\right\} = \int_0^\infty \left[e^{-\frac{\gamma_p}{\lambda_{11} P_p}} - e^{-\gamma_p \left(\frac{1+P_s \lambda_{21} y}{\lambda_{11} P_p}\right)}\right] e^{-y} dy \qquad (20)$$

$$P\{S_4\} = P\left\{\frac{P_p g_{11}}{P_s g_{21} + 1} < \frac{\gamma_p}{3}, P_p g_{11} > \gamma_p\right\} = \int_{\frac{2}{P_s \lambda_{21}}}^\infty \left[e^{-\frac{\gamma_p}{\lambda_{11} P_p}} - e^{-\frac{\gamma_p}{3}\left(\frac{1+P_s \lambda_{21} y}{\lambda_{11} P_p}\right)}\right] e^{-y} dy \qquad (21)$$

For calculation purposes, the region S_3' is defined as the union of S_3 and S_4.

$$S_3' = S_3 \cup S_4 \qquad (22)$$

$$P\{S_3'\} = P\left\{P_p g_{11} \geq \gamma_p, \frac{P_p g_{11}}{P_s g_{21} + 1} < \frac{\gamma_p}{2}\right\} = \int_{\frac{1}{P_s \lambda_{21}}}^\infty \left[e^{-\frac{\gamma_p}{\lambda_{11} P_p}} - e^{-\frac{\gamma_p}{2}\left(\frac{1+P_s \lambda_{21} y}{\lambda_{11} P_p}\right)}\right] e^{-y} dy \qquad (23)$$

$$P\{S_3\} = P\left\{S_3'\right\} - P\{S_4\} \qquad (24)$$

$$P\{S_2\} = P\left\{S_2'\right\} - P\{S_3\} - P\{S_4\} \qquad (25)$$

The region S_5' is also defined as the union of S_5, S_6 and S_7.

$$S_5' = S_5 \cup S_6 \cup S_7 \qquad (26)$$

$$P\left\{S_5'\right\} = P\left\{\frac{\gamma_p}{2} < P_p g_{11} < \gamma_p\right\} = e^{-\frac{\gamma_p}{2\lambda_{11} P_p}} - e^{-\frac{\gamma_p}{\lambda_{11} P_p}} \qquad (27)$$

$$P\left\{S_6'\right\} = P\left\{\frac{\gamma_p}{2P_p} < g_{11} < \frac{\gamma_p}{P_p}\left(\frac{P_s g_{21} + 1}{P_s g_{21} + 2}\right)\right\} = \int_0^\infty \left[e^{-\frac{\gamma_p}{2\lambda_{11} P_p}} - e^{-\frac{\gamma_p}{\lambda_{11} P_p}\left(\frac{P_s \lambda_{21} y + 1}{P_s \lambda_{21} y + 2}\right)}\right] e^{-y} dy \qquad (28)$$

$$P\{S_5\} = P\left\{S_5'\right\} - P\left\{S_6'\right\} \qquad (29)$$

$$\{S_7\} = P\left\{\frac{\gamma_p}{2P_p} < g_{11} < \frac{\gamma_p}{P_p}\left(\frac{P_s g_{21} + 1}{P_s g_{21} + 3}\right)\right\} = \int_{\frac{1}{P_s \lambda_{21}}}^\infty \left[e^{-\frac{\gamma_p}{2\lambda_{11} P_p}} - e^{-\frac{\gamma_p}{\lambda_{11} P_p}\left(\frac{P_s \lambda_{21} y + 1}{P_s \lambda_{21} y + 3}\right)}\right] e^{-y} dy \qquad (30)$$

The regions S_8' is the union of S_8 and S_9.

$$S_8' = S_8 \cup S_9 \qquad (31)$$

$$P\left\{S_8'\right\} = P\left\{\frac{\gamma_p}{3} < g_{11} P_p < \frac{\gamma_p}{2}\right\} = e^{-\frac{\gamma_p}{3\lambda_{11} P_p}} - e^{-\frac{\gamma_p}{2\lambda_{11} P_p}} \qquad (32)$$

$$P\{S_9\} = P\left\{\frac{\gamma_p}{3P_p} < g_{11} < \frac{\gamma_p}{P_p}\left(\frac{P_s g_{21} + 1}{2P_s g_{21} + 3}\right)\right\}$$

$$= \int_0^\infty \left[e^{-\frac{\gamma_p}{3\lambda_{11}P_p}} - e^{-\frac{\gamma_p}{\lambda_{11}P_p}\left(\frac{P_s\lambda_{21}y+1}{2P_s\lambda_{21}y+3}\right)} \right] e^{-y} dy \tag{33}$$

$$P\{S_8\} = P\left\{S_8'\right\} - P\{S_9\} \tag{34}$$

$$P\{S_{10}\} = P\left\{g_{11}P_p < \frac{\gamma_p}{3}\right\} = 1 - e^{-\frac{\gamma_p}{3\lambda_{11}P_p}} \tag{35}$$

5.2 Outage Probability Analysis

As explained in Sect. 3, there is no primary outage in regions S_1-S_9, while in S_{10}, primary transmission cannot succeed at any time. Accordingly, the outage probabilities of primary user for both *aggressive* and *conservative* SHARP are the same and equal to $P\{S_{10}\}$, as given by (34).

The corresponding outage probability for the secondary (P^{OS}) is given as:

$$P^{OS} = 1 - e^{-\frac{\gamma_s}{\lambda_{22}P_s}} \tag{36}$$

5.3 Throughput Analysis

The effective throughputs of the *aggressive* and *conservative* SHARP are dependent on which region the primary and secondary users are working due to the different transmission opportunities in each case. The throughputs of the primary and secondary users for the *aggressive* SHARP are:

$$G_p^A = R_p P\{S_1\} + \frac{R_p}{2}(P\{S_2\} + P\{S_5\})$$
$$+ \frac{R_p}{3}(P\{S_3\} + P\{S_4\} + P\{S_6\} + P\{S_7\} + P\{S_8\} + P\{S_9\}) \tag{37}$$

$$G_s^A = [R_s(P\{S_1\} + P\{S_2\} + P\{S_3\} + P\{S_{10}\}) + \frac{2R_s}{3}(P\{S_4\} + P\{S_6\})$$
$$+ \frac{R_s}{2}(P\{S_5\})\frac{R_s}{3}(P\{S_7\} + P\{S_8\})](1 - P^{OS}) \tag{38}$$

The *conservative* SHARP throughput is:

$$G_p^C = R_p\big(P\{S_1\} + P\{S_2'\}\big) + \frac{R_p}{2}\big(P\{S_5\} + P\{S_6'\}\big) + \frac{R_p}{3}\big(P\{S_8\} + P\{S_9\}\big)$$
$$\tag{39}$$

$$G_s^C = [R_s(P\{S_1\} + P\{S_{10}\}) + \frac{R_s}{2}(P\{S_5\}) + \frac{R_s}{3}(P\{S_8\})](1 - P^{OS}) \tag{40}$$

Since there are three possible transmissions being considered, more terms appear in the throughput analysis due to the reduction in the effective data transmitted.

6 Numerical Results

Numerical results of the proposed SHARP schemes are provided in this section. All channels follow the Rayleigh distribution. By default, the constant power for PU and SU, as well as the Gaussian background noise variance N, are equal to unity.

Figure 4 contains the achievable throughput for the *conservative* and *aggressive* SHARP considering both cases of two [7] and three retransmissions by varying the secondary transmitter power. The curves did not change significantly by adding a retransmission. However, the primary user experienced a gain in performance for the conservative scheme, opposite of secondary, whereas in the aggressive scheme, there is a tradeoff between primary and secondary depending on the secondary transmission power. This result shows that the system may generally have this behavior regardless the number of transmissions.

Figure 5 provides the throughput evolution with the signal-to-interference ratio (SIR) at the primary receiver, keeping the same transmission power for both PU

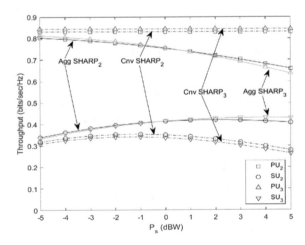

Fig. 4 Throughput comparison versus P_s $\big[R_s = 0.5, R_p = P_p = N = 1, \lambda_{11} = \lambda_{22} = 4, \lambda_{21} = 1\big]$

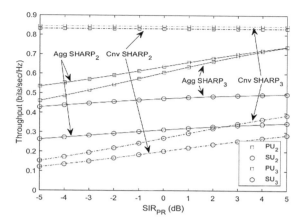

Fig. 5 Throughput comparison vs. $SIR_{PR}[R_p = 1, P_p = P_s = N = 1, \lambda_{11} = 4, \lambda_{22} = 1]$

and SU, constant primary and secondary link channel gain and varying the interfering channel between them. The *conservative* SHARP improves the throughput considerably with higher transmissions. However, the same does not occur with the *aggressive* SHARP, which may suffer in high interference or low power scenarios. This may be explained by the greater interference caused by the secondary with more retransmissions. Generally, the throughput may increase with a higher SIR.

7 Conclusion

This paper aims an extension in the study of the proposed scheme for underlay cognitive radio named SHARP [7] by considering a greater number of retransmissions. Same reasoning was applied to find all the operating Regions and their respective probabilities, as long as the discovery mechanism and flow chart representation. In general, comparing all results with the ones obtained in [7], it may be noticed that the number of retransmissions do not change the shape of the curves in the figures, indicating a general behavior. Nevertheless, more possibilities arise from a higher number of retransmissions, as illustrated in the flow charts presented before. Numerical results show that the *conservative* SHARP may increase performance directly to the number of retransmissions, whereas the *aggressive* SHARP improves its operation in low power scenarios. By considering the generalization with n transmissions, the curves may follow the same behavior without significant changes but a slightly gain in performance. The trade-off between performance gain and higher latency in retransmissions must be taken into account. Future studies may approach the latency influence depending on the number of retransmissions.

References

1. Haykin S (2005) Cognitive radio: brain-empowered wireless communications. IEEE J Sel Areas Commun 23(2):201–220
2. Zhao Q, Sadler BM (2007) A survey of dynamic spectrum access. IEEE Signal Process Mag 24(3):79–89
3. Chandwani N, Jain A, Vyava PD (2015) Throughput comparison for cognitive radio network under various conditions of primary user and channel noise signals. In: 2015 Radio and Antenna Days of the Indian Ocean (RADIO). IEEE, pp 1–2
4. Park C-H, Kim S-W, Lim S-M, Song M-S (2007) HMM based channel status predictor for cognitive radio. In: 2007 Asia-Pacific microwave conference. IEEE, pp 1–4
5. Barnawi A (2010) A novel approach for cognitive radio sensing using wideband chirp signal. In: IEEE international microwave workshop series on RF front-ends for software defined and cognitive radio solutions (IMWS). IEEE, pp 1–4
6. Hattab G (2014) Multiband spectrum access: great promises for future cognitive radio. Proc IEEE 102(3):282–306
7. Li JCF, Zhang W, Nosratinia A, Yuan J (2013) SHARP: spectrum harvesting with ARQ retransmission and probing in cognitive radio. IEEE Trans Commun 61(3):951–960
8. Bahgat MM (2019) Design of low cost cognitive radio platform for demonstration and testing purposes. In: IEEE international conference on design & test of integrated micro & nano-systems (DTS), pp 1–5
9. Roy S, Fortier P (2004) Maximal-ratio combining architectures and performance with channel estimation based on a training sequence. IEEE Trans Wireless Commun 3(4):1154–1164

Impact on Compressed Sensing for IoT Used Indoor Environment Monitoring System

Nobuyoshi Komuro and Reiji Suzumaru

Abstract This paper proposes a method to save power consumption for Wireless Sensor Network (WSN) based environment monitoring system by using compressed sensing technique. Reconstruction process of compressed sensing data is complicated but compressed process is itself simple. On the other hand, although sensor nodes can use limited resources, the server has affluent resources. Therefore, sensor nodes compress time series measured environment data and the server reconstruct the compressed data. At the environment data collection stage, a sensor node uses compressed sensing for transmitting measured environment data. The server reconstructs the received environment data. This study develops the ZigBee WSN based time series indoor environment data collection system. Then this study investigates the impact on compressed sensing technique for WSN. Experiment results show that when setting compression ratio = 40%, the power consumption was reduced by 40% with satisfying the coefficient of determination > 0.8, which showed the effectiveness of the proposed method.

Keywords Wireless Sensor Network (WSN) · Internet of Things (IoT) · Compressed sensing · Sparse modeling · Power saving · Environment monitoring

1 Introduction

In recent years, Ad-hoc Networks and Wireless Sensor Networks (WSNs) gain enormous attention by researchers with the diversification of the applications, such as medical applications and environmental monitoring systems [1–12]. In order to implement WSN to various applications, it is important to save power consumption of wireless sensor devices.

The IEEE 802.15.4 [13] is the standard Medium Access Control (MAC) protocol for Low-Rate Wireless Personal Area Networks (LR-WPANs). In the IEEE 802.15.4 standard, the packet transmission is conducted under the superframe structure. The

N. Komuro (✉) · R. Suzumaru
Chiba University, 1-33 Yayoi-cho, Inage-ku, Chiba-shi, Chiba 263-8522, Japan
e-mail: kmr@faculty.chiba-u.jp

© The Author(s), under exclusive license to Springer Nature Singapore Pte Ltd. 2021
H. Kim et al. (eds.), *IT Convergence and Security*, Lecture Notes
in Electrical Engineering 712, https://doi.org/10.1007/978-981-15-9354-3_9

IEEE802.15.4 compliant devices manage the power consumption by controlling the active and inactive cycles (Duty-Cycle). Duty-Cycle are controlled to reduce idle listening causing the power waste in the IEEE 802.15.4 standard.

MAC protocols for achieving low power WSN have been proposed [14–17]. MAC protocols in [16, 17] tune MAC layer parameter and duty-cycle in order to achieve the required QoS. These MAC protocols reduce data transmission opportunity in order to save power consumption. Since WSN often collects time series data, reduction of the transmission opportunities may have a bad influence on application systems.

On the other hand, compressed sensing techniques have been investigated for reconstructing high resolution images [18–20]. According to the compressed sensing theory, a sparse signal can be accurately reconstructed with a relatively small number of measurement. Since compressing process is simply performed without introducing excessive computational and control overhead, compressed sensing technique may be suitable for WSN.

This paper proposes a method to save power consumption for WSN based environment monitoring system by using compressed sensing technique. Reconstruction process of compressed sensing data is complicated but compressed process is itself simple. On the other hand, though sensor nodes can use limited resources, the server has affluent resources. Therefore, a sensor node compresses time series measured environment data and the server reconstruct the compressed data. At the environment data collection stage, a sensor node uses compressed sensing for transmitting measured environment data. The server reconstructs the received environment data. First this study develops the ZigBee WSN based time series indoor environment data collection system. Then this study investigates the impact on compressed sensing technique for WSN. Finally, this study evaluates the effectiveness of the proposed system by experiment.

2   System Structure

The proposed system collects and saves indoor environment data. Sensor node measures environment data. A sensor node compresses the measured data and sends the compressed data to the server. The server reconstructs the received data from sensor node. A sensor node compresses the measured data by multiplying random matrix which is composed of 0 or 1. A sensor nodes work with a light load. In addition, a sensor node can reduce data transmission number. Therefore, it is expected for a sensor node to save power to transmit the measurement data. On the other hand, compared to the compression process, data reconstruction process is complicated. Since the server has affluent resources, the reconstructed process is done at the server.

2.1 Sensor-Operation Control

In order to develop measurement equipment, it is necessary to control sensors' operations. Each sensor's operation is controlled by on-board microcomputer. The proposed system uses Arduino as a one-board microcomputer. Each sensor measures the indoor environment data periodically. In this system, sensors measure the temperature, humidity, and illuminance. Sensor nodes send the measured data through a wireless module.

2.2 Network Construction

This system uses XBee as a wireless module. A star topology sensor network is constructed. Each sensor sends the measured data to the coordinator node. The coordinator node transfers the received data to the server through serial communication. The sever saves the collected data as csv files. The files include measured data, sensor ID, and the sensor data reception time.

2.3 Sensor Node

This work developed indoor environment sensor which include temperature and humidity sensors, and illuminance sensors. The environment sensor gets output voltages according to the temperature, humidity, and illuminance around the sensors. The one-board microcomputer converts the obtained voltages to corresponding temperature (Degree Celsius), humidity (%), and illuminance (LUX).

2.4 Compressing Algorithm

Let n be the number of sensor data, m be the number of compressed sensor data, $x \in \mathbb{R}^n$ be measured sensor data, and Φ be $m \times n$ observation matrix. Then the compressed sensor data is expresses as

$$\mathbf{y} = \Phi \mathbf{x}$$

As an observation matrix, a sparse random matrix, which is composed of 0 or 1, is used. The example of the relationship between the observation matrix and sensor data is shown in Fig. 1.

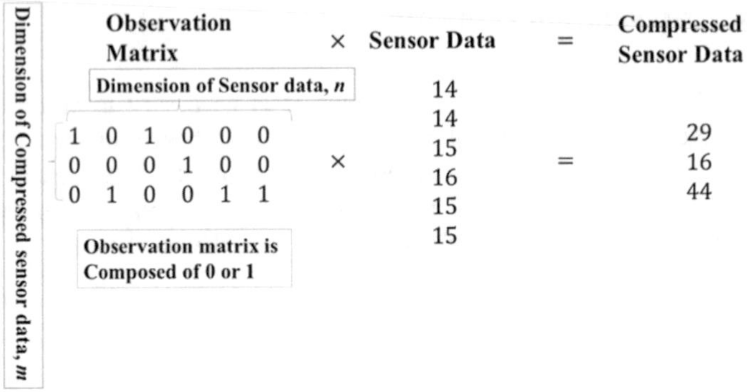

Fig. 1 Example of relationship between observation matrix and sensor data

2.5 Reconstruction Algorithm

Let $s \in R^n$ be sparse representation of original data x, and Ψ be $n \times n$ base transformation matrix. The relationship between the original sensor data and the sparse data is expresses as

$$x = \Psi$$

In order to represent sparse signal, a base transformation matrix is used. The observation matrix and the base transformation matrix must be incoherent. This paper uses the Discrete Cosine Transform (DCT) based base transformation matrix. Let $\phi_k[i]$ be DCT basic. DCT basic Ψ for i-th column and k-th line is defines as

$$\phi_k[i] = \begin{cases} \frac{1}{\sqrt{N}} & (k = 0) \\ \sqrt{\frac{2}{N}} \cos \frac{(2i+1)k\pi}{2N} & (k = 1, 2, ..., N - 1), \end{cases}$$

where N is the number of lines. The sparse data can be reconstructed by solving an L1 norm optimization problem, which is expressed as

$$\arg \min \|\hat{s}\| \, subject \, to \, \Phi\Psi\hat{s} = y.$$

2.5.1 Experiment

There were one temperature sensor node and one coordinator node in the network. The sensor node measured the temperature every 2 min. Sensor data were logged for 5 days. The sensor node compressed the temperature data and sent the compressed data. The coordinator transferred the received data from the sensor node and saved

Table 1 Sensor node parameters

Parameter	Unit	Value
Operating voltage	V	5.00
Consumed current in active mode	mA	16.26
Consumed current in sleep mode	mA	0.0657
A packet size	byte	127
Data rate	kbps	9.6

them on the server. The server reconstructed the compressed data. Parameters of a sensor node is shown in Table 1.

Figure 2a–i shows the measured sensor data and reconstructed data when the compressed ratios are 10%, 30%, 40%, 50%, 70%, and 90%, respectively. It is seen from Fig. 2 that the cross correlation between the measured sensor data and the reconstructed data depends on the compression ratios. Although the cross correlation between them is low at the high compression ratios, such as 70 and 90%, the cross correlation increases according to the decrease in the compression ratios. The reconstructed data shows relatively similar shape to the measured sensor data when the compression ratio is equal to or less than 50%. Also it is seen from Fig. 2 that the reconstructed data shows similar shape to the measured sensor data when the compression ratio is equal to or less than 40%.

Figure 3 shows the coefficient of determination R^2 and the power consumption reduction ratio as a function of the compression ratio. The green line shows R^2 and the blue line shows the power consumption reduction ratio. The coefficient of determination R^2 is expressed as

$$R^2 = 1 - \frac{\sum_{i=1}^{N} \left(y_i - \widehat{y}_i \right)^2}{\sum_{i=1}^{N} \left(y_i - \overline{y} \right)^2},$$

where N is the number of data, y is the original data, $\overline{y}$ is the average of the original data, and $\widehat{y}_i$ is the reconstructed data. Assuming that the power consumption in the sleep mode can be neglected, the power consumption reduction ratio E is expressed as

$$E = 1 - \frac{n^{(CS)} V I_a}{n V I_a}$$

where n is the number of transmissions of a sensor node without compressed sensing, $n^{(CS)}$ is that with compressed sensing, V (5 V) is the operating voltage, and I_a (=16.26 mA) is the consumed current in the active mode.

It is seen from Fig. 3 that the power consumption reduction ratio increases proportionally to the compression ratio. It is also seen from Fig. 3 that the compression ratio

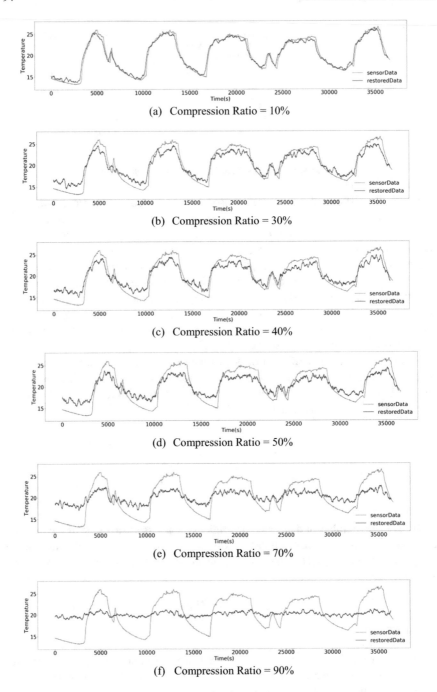

(a) Compression Ratio = 10%

(b) Compression Ratio = 30%

(c) Compression Ratio = 40%

(d) Compression Ratio = 50%

(e) Compression Ratio = 70%

(f) Compression Ratio = 90%

Fig. 2 Measured sensor data and reconstructed data in time

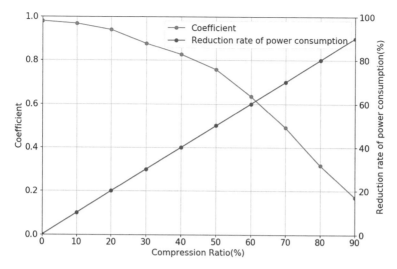

Fig. 3 Coefficient of determination and power consumption reduction ratio as a function of compression ratio

should be set to 40% for satisfying $R^2 \geq 0.8$. Similarly, the compression ratio should be set to 50% for satisfying $R^2 \geq 0.7$.

3 Conclusion

This paper proposed a method to save power consumption for WSN based environment monitoring system by using compressed sensing technique. In the proposed method, a sensor node compresses time series measured environment data and the server reconstruct the compressed data. At the environment data collection stage, a sensor node uses compressed sensing for transmitting measured environment data. The server reconstructs the received environment data. We developed the WSN based time series indoor environment data collection system. Then we investigated the impact on compressed sensing technique for WSN. Experiment results showed that when setting compression ratio = 40%, the power consumption was reduced by 40% with satisfying the coefficient of determination > 0.8, which showed the effectiveness of the proposed method. Future work includes the implementing compressed sensing mechanism into multiple sensor nodes.

References

1. Tseng SM (2003) A high-throughput multicarrier DS CDMA/ALOHA network. IEICE Trans Commun E86-B(4):1265–1273
2. Komuro N, Habuchi H, Kamada M (2004) CSK/SSMA ALOHA system with nonorthogonal sequences. IEICE Trans Fundam E87-A(10):2564–2570
3. Komuro N, Habuchi H (2005) A reasonable throughput analysis of the CSK/SSMA unslotted ALOHA system with nonorthogonal sequences. E88-A(6):1462–1468
4. Sekiya H, Tsuchiya Y, Komuro N, Sakata S (2011) Analytical expression of maximum throughput for long-frame communications in one-way string wireless multihop networks. Wireless Pers Commun 60(1):29041
5. Komuro N, Habuchi H, Tsuboi T (2008) Nonorthogonal CSK/CDMA with received-power adptive access control scheme. IEICE Trans Fundam E91-A(10):2779–2786
6. Kobayashi K, Nakamura A, Ohno K, Itami M (2016) Improving performance of DS/SS-IVC scheme based on location oriented PN code allocation. IEICE Trans Fundam E99-A(1):225–234
7. Heinzelman WR, Chandrakasan A, Balakrishnan H (2000) Energy-efficient communication protocol for wireless microsensor networks. In: Proceedings of annual Hawaii international conference on system sciences, pp 223–232
8. Lindsey S, Raghavendra C, Sivalingam KM (2002) Data gathering algorithms in sensor networks using energy metrics. IEEE Trans Parallel Distrib Syst 13(9):924–935
9. Fan X, Song Y (2007) Improvement on LEACH protocol of wireless sensor network. In: Proceedings of international conference on sensor technologies and applications (SENSOR-COMM), pp 260–264
10. Luo CY, Komuro N, Takahashi K, Kasai H, Ueda H, Tsuboi T (2008) Enhancing QoS provision by priority scheduling with interference drop scheme in multi-hop ad hoc network. In: IEEE global communication conference (GLOBECOM), pp 1321–1325
11. Tuan DT, Sakata S, Komuro N (2012) Priority and admission control for assuring quality of I2V emergency services in VANETs integrated with wireless LAN mesh networks. In: International conference on communications and electronics (ICCE), pp 91–96
12. Sony CT, Sangeetha CP, Suriyakala CD (2015) Multi-hop LEACH protocol with modified cluster head selection and TDMA schedule for wireless sensor networks. In: Proceedings of global conference on communication technologies (GCCT), pp 539–543
13. IEEE Standards Association (2011) IEEE standard for local and metropolitan area networks Part 15.4: low-rate wireless personal area networks (LR-WPANs). https://ecee.colorado.edu/~liue/teaching/comm-standards/2015S-zigbee/802.15.4-2011.pdf
14. Lee B-H, Wu H-K (2010) Study on a dynamic superframe adjustment algorithm for IEEE802.15.4 LR-WPAN. In: Proceedings of IEEE VTC. Spring
15. Sun Y, Gurewitz O, Johnson DB (2008) RI-MAC: a receiver-initiated asynchronous dutycycle MAC protocol for dynamic traffic loads in wireless sensor networks. In: Proceedings of 6th ACM conference on embedded network sensor systems, pp 1–14
16. Kai N, Sakata S, Komuro N (2013) Dynamic active period control achieving low energy consumption and low latency in multi-hop wireless sensor networks. In: IEEE eleventh international symposium on autonomous decentralized systems
17. Akbar MS, Yu H, Cang S (2017) TMP: tele-medicine protocol for slotted 802.15.4 with duty-cycle optimization in wireless body area sensor networks. IEEE Sens J 17(6):1925–1936
18. Adcock B, Hansen AC (2016) generalized sampling and infinite-dimensionalcompressed sensing. Found Comput Math 16:1263–1323
19. Feng L, Axel L, Chandarana H, Block KT, Sodickson DK, Otazo R (2016) XD-GRASP: golden-angle radial MRI with reconstruction of extra motion-state dimensions using compressed sensing. Magn Reson Med 75:755–788
20. Beygi S, Jalali S, Maleki A, Mitra U (2019) An efficient algorithm for compression-based compressed sensing. Inf Inference J IMA 8(2):343–375

Development and Performance Testing of the Automated Building Energy Management System with IoT (ABEMS-IoT) Case Study: Big-Scale Automobile Factory

Nichakul Imtem, Chatchai Sirisamphanwong, and Nipon Ketjoy

Abstract This article presented the development of the automated building energy management system using the Internet of Things (IoT) so-called aBEMS-IoT, and it was implemented in the Big-Scale Automobile Factory in Thailand. The aBEMS-IoT consisted of three parts, hardware which consists of the smart sensors and control devices, software which compost of intelligent control algorithm, report, mobile application, and notification by application LINE, a communication system is using a wireless solution (LoRa). The communication and control devices are to enable Plug & Play installation and set various parameters through the mobile phone. Moreover, the aBEMS-IoT connect a database and show the information in a graphical interface. The results indicated that the aBEMS-IoT, which was installed in the automobile factory work properly and was able to save the money of 12%. This system will be the technology for supporting demand response in the future.

Keywords Building energy management · Demand-side management · Smart building · Demand response

1 Introduction

At present, every industry is faced with technology disruption because advanced technology is used to replace human labor. For this reason, systematic planning is needed for the near future development. Not even, the energy industry has to adapt to coming disruption technology [1–4]. During the daytime, a high amount of electrical energy is injected into the grid by solar systems, but during nighttime, the peak load appears; this pattern may create the duck curve. This case may negatively influence

N. Imtem · N. Ketjoy
School of Renewable Energy Technology and Smart Grid Technology, Naresuan University, Phitsanulok 65000, Thailand

C. Sirisamphanwong (✉)
Smart Energy System Integration Research Unit, Department of Physics, Faculty of Sciences, Naresuan University, Phitsanulok 65000, Thailand
e-mail: chatchaisi@nu.ac.th

H. Kim et al. (eds.), *IT Convergence and Security*, Lecture Notes in Electrical Engineering 712, https://doi.org/10.1007/978-981-15-9354-3_10

the power quality, voltage, and frequency fluctuations of the grid network. The flex-ible load profile is one of the methods that can flatten the duck curve by controlling the power of both sites, supply, and demand. The demand response (DR) is to change in load profile by the user from the normal load power pattern in response to grid stability, incentive payment, or electricity prices. The techniques of the DR consist of load building, load shifting, valley filling, peak clipping, and load conservation [5–10]. Demand-side management (DSM) is significant for smart grid development; hence the advance technology for managing load demand has been developed, such as BEMS, FEMS, and HEMS [8–10]. This research focused on the development of the automated building energy management system using IoT called the aBEMS-IoT. This purpose is to develop a building energy management system (BEMs) for controlling the load demand follow the signal control from the load aggregator or energy pricing. The system will work automatically and easy to control response to various situations, over and under grid voltage or responding to the electrical pricing. The idea of developing a BEMS-IoT is using a wireless data transmission device that covers a wide area, including closed environments such as tall buildings or the area which is challenging to deal with wire. Not only closed environments but also open environments such as a factory area with a wide-open space is also a good deal with wireless data transmission, which can be easily installed, used conveniently, and have a reasonable cost worth the investment. For supporting the DR programs, the buildings which are installed aBEMS-IoT, they are upgraded to smart building. The aBEMS-IoT consists of three parts as follows. The first is hardware, measuring, controlling devices which enable to manage the load On/Off, increase/decrease the power demand of the user. The second is software, which is a data processing, monitoring system which is analyzing the data from both sides load aggregator and customers then after data analysis sends the signal to the control devices for the future process. The third is the communication system. After aBEMS-IoT is developed and implemented in the Big-Scale Automobile Factory, it will be tested to find out the technical issue such as working principle, money-saving when using the aBEMS-IoT response to the TOU pricing.

2   The ABEMS-IoT Development

The automated building energy management system using IoT (aBEMS-IoT) was developed for supporting the DR program to maintain the grid network stability and energy saving on the user side. The purpose of the aBEMS-IoT is to control or manage the load power automatically, that response to the various situations, the requirement of the load aggregator in case of an emergency, or may respond to electrical pricing for money-saving of the user side. The product enables us to measure all concerning parameters and also can operate for demand-side management. It can send and receive data on many devices via Modbus protocol, then transfer data through the LoRaWAN communication network and send data to the central processing system for fast and accurate power consumption analysis and decision. The concept of the aBEMS-IoT is to manage the load power in the building by Edge analytics and control, its sending and receiving the signal through the IoT Gateways [11–13]. The communication between the IoT Gateways and load can be the wireless, cable, or using both of them. The aBEMS-IoT components can be classified into three main parts; the first is hardware, the second is software, and the third is communication. The details of all components can be described as following.

2.1 The Control System (Hardware)

The control system (hardware) which enables to control the load On/Off, increase/decrease the power consumption of the load demand, marking it flexible. The loads which were controlled by the aBEMS-IoT consisted of the pump, water sprinkler, air conditioner, lighting, plug, environment monitoring. These main components of this part consist of MicroController, which was developed by our team (Fig. 1).

2.2 The Data Analytics and Monitoring System (Firmware)

The firmware was developed for energy management and monitoring system in the smart building for demand-side management. This firmware is possible to operate automatically control the load or any appliances following the requirements of users or load aggregators. These parts will consist of energy monitoring and data analytics for smart demand response to any situation and requirements. This firmware will measure all concerning parameters and control the electrical appearances such as pump, water sprinkler, air conditioner, lighting, plug, parking car, refrigerator, and beakers via the internet. An edge gateway can connect smaller mesh sensor networks and non-Ethernet/IP-based devices (e.g., serial BACnet* and Modbus*) to an IT network, data center, or cloud [6, 14–16]. Gateways support hybrid configurations of wired and wireless sensor connections. For example, a gateway measuring power with a wired current clamp at the main panel can also collect data from nearby wireless mesh sensors measuring voltage and current. Data analytics can be centralized in a data center or cloud, decentralized in edge devices (e.g., gateways), or a combination of the two. Centralized data analytics is relatively easy to implement and maintain since the software runs in one place; however, this approach could require a significant amount of network bandwidth to carry as much as a thousand data points per second. Using a cellular network to send large amounts of building data to the cloud could be rather costly. Backhauling all of this data to a complete centralized analytics structure also requires a substantial amount of storage in the location where it is most expensive [17, 18] (Fig. 2).

The multiple data sources (Real-time, structured, and unstructured data) is collected and sent via the IoT gateway to the data analytics part. After the data is analyzed, the control signal will be sent to load via the IoT gateways for controlling

Fig. 1 The control and communication bord of the a BEMS-IoT

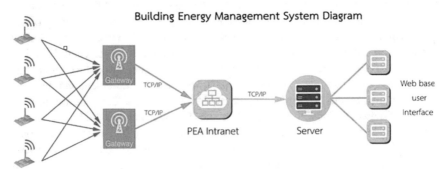

Fig. 2 The data analytics and monitoring system

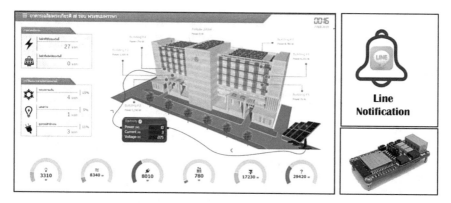

Fig. 3 The monitoring system of the aBEMS-IoT

the load as following the price of electricity or emergency. The result of the process will be reported to the user via the monitoring system, which can be accessed via the mobile, dashboard, and other internet devices. The data analytics and monitoring system of the aBEMS-IoT is presented in Fig. 3.

3 The Implementation of ABEMS-IoT at a Big-Scale Automobile Factory

The aBEMS-IoT was implemented in the big-scale automobile factory in Thailand; many advance technology is installed, such as measurement devices (Power meter, temperature sensor, control unit at the appliance devices. Not only was the demand-side managed but also the supply-side, PV rooftop, is also monitored and controlled. A big-scale automobile factory is presented in Fig. 4, and the architecture is presented in Fig. 5.

A automate building energy management (aBEMS-IoT) is a cost-effective computer-based control system installed in buildings of Automobile Factory that monitors and controls the building's mechanical and electrical equipment such as air-conditioners, lighting, power systems, parking systems, and security systems. This system is a complete intelligent design management and control system that consists of combines hardware and software for controlling the load demand of any building in a factory. Hardware is set up by many sensors, temperature, humidity, and control of the power system; meanwhile, the software integrates the database collection, monitoring, control algorithm. The main purpose of this system is to make a flexible load profile for any situation, response to an emergency or electrical pricing (TOU). The aBEMS-IoT enables us to complete in four basic functions. The first is a monitoring system which is monitoring all sensors device to the behavior of energy use, and the second is a controlling system which controls the electrical equipment or any power system to decrease/increase in load power, the third is optimizing which is

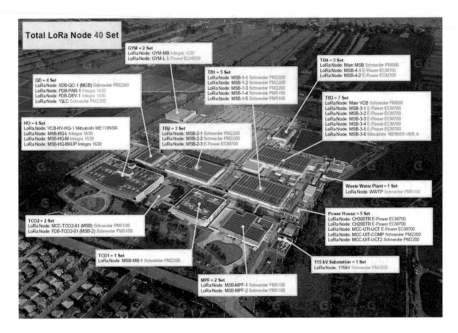

Fig. 4 The aBEMS-IoT at big-scale automobile factory

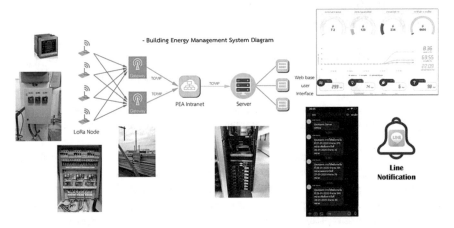

Fig. 5 The architecture of an aBEMS-IoT at big-scale automobile factory

analyzing the data from all devices then a decision for next process, finally, the forth is reporting which is report all final result to the operators. The advantages of this system are the following; smart data collector communicates with various devices in the factory used to measure or control power system by reading and executing via Modbus protocol. The communication and control devices are to enable Plug &

Play installation, and it can set various parameters through the mobile phone. Moreover, conceptually, the aBEMS-IoT connect a database and show the information in a graphical interface in the form of a final report to the operator. Finally, the high-level algorithms are included within this system for helping the users, save-money by operation following the TOU structure and also maintain the power quality in the factory.

4   Validation of a BEMS-IoT

In this section, the evaluation of the developed the aBEMS- IoT which was installed in the Big-Scale Automobile Factory in Thailand from different perspectives.

4.1   Data Monitoring

The aBEMS-Iot of the Automobile Factory enables full monitoring of data from the wireless sensor networks (LoRa) of all areas which are installed the Power meters in the factory. An example of the data monitored is presented in Fig. 6.

Figure 6, presents the ratio of different types of electrical appliances and power system, which consist of air conditioners, evaluators, lighting, electrical appliances in the office, and the other electrical devices. According to Fig. 8, found that the air-conditioner consumed 40% of electrical energy, the lighting system used 15, 5% was supplied to the appliances in the office, and the rest of the electrical energy was supplied to another power system. With all results, it is possible to assess which load types can be provided for demand response. The aBEMS-IoT also enables the monitoring of activities within the Automobile Factory, which analyzed which activities were not necessary (Fig. 7).

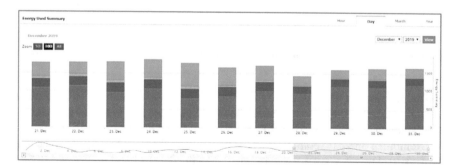

Fig. 6 The summary of daily energy consumption

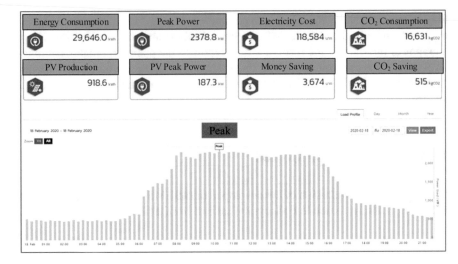

Fig. 7 Shows the stairwell usage of the testbed for

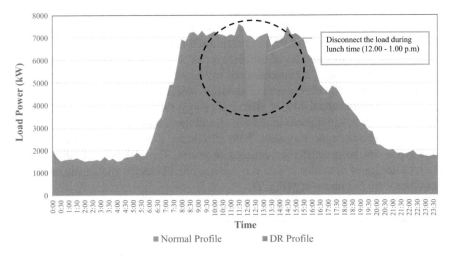

Fig. 8 Comparison between normal and DR load profile

The software of the aBEMS-IoT in the factory was developed to generate a summary report to the operator automatically. It was used in a field that purposed at creating demand response for saving money, power quality, and electrical stability. An example of one-day monitoring on 2020–02-18 as shown in Fig. 8, it showed that the peak load was 2,379 kW appeared at 10.30 a.m., the electrical price was 118,584 THB/day (3,825 USD/day), and the electricity production from PV system was 919 kWh/day. Not only the energy and financial information but also environmental data were presented as shown the CO_2 saving (515 kg/day). The report is automatically generated every 24 h (one day), and it contained hourly, daily, monthly, and yearly final analyzed information that contained electricity usage at the appliance level and also showed the consumption of the different appliance in comparison with other sections. The executive board used these reports as a baseline for discussion on energy consumption behavior and can be used to gain insights into which appliances can provide demand response in case if required.

4.2 Load Control

The aBEMS-IoT supports the direct and indirect control process after the system collected the information from sensing devices and analyzed them, then the next process is making the decision for controlling the controllable load in the factory. An example of a control setpoint of the air-conditioner can be dived into two modes, the first is operation mode which is out of the range 24–26 °C and the second is standby mode which is in the range of 24–26 °C. This algorithm is simple, but it is still proper operation, but in the future, the advanced algorithm will be developed using the Fuzzy Logic Based. The factory was occupied during the intervention, and the indoor climate conditions were checked to ensure occupants' comfort and fulfillment of the factory. The aBEMS-IoT is able to set the time for start/stop of the controllable appliances and power system when it is not necessary to use such as 40% of air-conditioners is set to stop during the lunchtime, some of the electrical machinery automatically stop when the worker is out of the working area and easy to understand is that lighting and appliances are in the working, all of them will automatically stop when the worker is not in the control area. Load control by the aBEM-IoT allows the factory to reduce the energy consumption and also can decrease the peak load (Peak Clipping) and possible to save the money by peak load shifting response to electricity price structure (TOU). The example of load control by the aBEMS-IoT in case of managing the air-conditioners and some controllable appearances during the lunchtime is presented in Figs. 8 and 9.

Figure 11, presented the group of the appliances which much consumed the electricity in the factory and analyzed them for the energy efficiency and the solution for decreasing the energy demand. The aBEMS-IoT can find the fault in their factory, such as peak power during off-day, error in the PV power plant or some of the power system is not turnoff and etc. This system sends the error message to the operator by line using line notification.

5 Conclusion

This research developed the automated building energy management system using the internet of thing so-called aBEMS-IoT and it was installed in the Big-Scale Automobile Factory of Thailand. This the aBEMS-IoT consisted of three parts, and

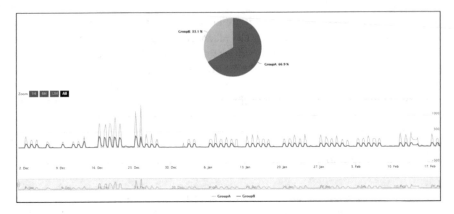

Fig. 9 Comparison between a group of appliances which consumed energy

the first is hardware which consists of the smart sensor and the smart control device, the second is software which compost of intelligent control algorithm, smart report and mobile application. The third is a communication system which is a wireless solution (LoRa). This system can be analyzed and supported by the different levels of interaction with all parts of the factory. The results of the research indicated that the aBEMS-IoT which was installed in the automobile factory is able to save the money of 12% compared to the system without the aBEMS-IoT. Furthermore, this system can be analyzed the ratio of the energy consumption per habitant, showed using energy efficiency. The aBEMS-IoT is multi-modal because demand response provision and local energy efficiency in the factory control are seamlessly supported. The relatively low demand response potential is consistent with the building being very modern and energy-efficient.

References

1. Minoli D, Sohraby K, Occhiogrosso B (2017) IoT considerations, requirements, and architectures for smart buildings—energy optimization and next-generation building management systems. IEEE Int Things J 4:269–283
2. Al-Ali A, Zualkernan I, Rashid M, Gupta R, Alikarar M (2017) A smart home energy management system using IoT and big data analytics approach. IEEE Trans Consum Electron 63(4):426–434
3. Martirano L et al (2017) Demand side management in microgrids for load control in nearly zero energy buildings. IEEE Trans Ind Appl 53:1769–1779
4. Majumdar S et al (2018) User-level runtime security auditing for the cloud. IEEE Transac Inf Forens Sec 13:1185–1199
5. Liu J et al (2017) A two-stage residential demand response framework for smart community with transformer aging. IEEE PES Asia-Pacific Power and Energy Engineering Conference (APPEEC), 1–6
6. Griful S, Welling U, Jacobsen R (2017) Implementation of a building energy management system for residential demand response. Microproc Microsyst 55:100–110

7. Ming J et al (2017) MOD-DR: microgrid optimal dispatch with demand response. Appl Energy 187:758–776
8. Mian H et al (2018) Distributed real-time demand response for energy management scheduling in smart grid. Int J Electr Power Energy Syst 99:233–245
9. Eissa M (2017) Developing incentive demand response with commercial energy management system (CEMS) based on the diffusion model, smart meters and new communication protocol. Appl Energy 236:273–292
10. Guangsheng P et al (2019) Optimal design and operation of multi-energy system with load aggregator considering nodal energy prices 239:280–295
11. Dongmin Y, Huanan l, Charis B (2018) Peak load management based on hybrid power generation and demand response. Energy 969:969–985
12. Brady S, Elizabeth B, Elaine H (2017) The value of demand response in Florida. Elect J 30:57–64
13. Ailin A et al (2018) Evaluation of residential customer elasticity for incentive-based demand response programs. Elect Power Sys Res 158:26–36
14. Ailin A et al (2017) A smart home energy management system using iot and big data analytics approach. IEEE Trans Consum Electron 63:426–434
15. Kang W, Xie L, Choi H (2018) Impact of data quality in home energy management system on distribution system state estimation. IEEE Access 6:11024–11037
16. Abogaleela et al (2017) Reliability evaluation framework considering OHL emergency loading and demand response. IEEE PES Innovative Smart Grid Technologies Conference Europe (ISGT-Europe), 1–6
17. Pérez L, Rodriguez J (2018) Simulation of scalability in IoT applications. International Conference on Information Networking (ICOIN), Chiang Mai, Thailand, 577–582
18. Sergi R, Ubbe W, Rune H (2017) Implementation of a building energy management system for residential demand response. Microproc Microsyst 55:100–110

Part III
Computer Vision and Applications

A Layered Framework for Virtual Guidance to Network Maintenance Based on Augmented Reality

Syed Akhter Hossain, Rishad Islam, Shahir Rahman, and Shahriar Nayeem

Abstract With the growing interest in Augmented Reality (AR) re-searchers and developers are engaged in developing systems with AR technologies. AR provides a unique experience in visualizing the solutions and provides an exciting way of inter-acting with the real world. The network management system is a complicated process that can be addressed efficiently by AR technologies. The visualization of network topology information of network devices decreases the burden of the network administrator to a great extent. In this paper, we proposed frameworks for both Platform Independent Model and Platform Specific Model which will be acting as a guide for the development of an application based on AR. We further developed a prototype for visualizing the network device information using the proposed framework. The prototype is acting as Proof of Principle (PoP) of our proposed framework. This prototype can be fully developed into a working application and installed in any organization with a network system as a guide for net-work device management and maintenance. The framework can also be modified to support various types of AR in different systems as per the requirement of the developer.

Keyword Augmented reality · Network device · Network maintenance · AR framework · Network server · Network administrator

S. A. Hossain
Military Institute of Science and Technology, Dhaka, Bangladesh
URL: https://mist.ac.bd/

R. Islam (✉) · S. Rahman · S. Nayeem
Department of Computer Science and Engineering, Daffodil International University, Dhaka, Bangladesh
e-mail: rishadxjcc@gmail.com
URL: http://daffodilvarsity.edu.bd/

1   Introduction

Nowadays, AR applications are becoming the backbone of the education industry. Apps are being developed which embed text, images, and videos, as well as real-world curriculum. Printing and advertising industries are developing apps to display digital content on top of real-world magazines [1]. Augmented reality displays super-imposed information in our field of view and can take us into a new world where the real and virtual worlds are tightly coupled. It is not only limited to desktop or mobile devices, different wearable computer devices like Google Glass, Microsoft HoloLens with an optical head-mounted display, are perfect examples of using AR technologies [2].

Network management is a vital part of the digital economy. They are fast-moving environments with high demanding reliability and availability requirements. Especially in the data center, there are several operations such as connecting cables, network connectivity, power distribution, storage units, etc. which are running all at once. Also, the system produces a huge amount of data from security tools and network monitoring tools [3]. Nowadays all applications and activities are based on providing service in the least response time. As AR applications are now gaining popularity in the industries for the ease of development we would like to incorporate this technology to manage the network designs and its maintenance.

The goal of this research is to bridge the fields of Augmented Reality and Network Maintenance by developing a framework along with a prototype and demonstrate the benefits of using an Augmented Reality interface in this particular field.

2   Related Work

In recent and past years, Several studies have been conducted on network de-vice identification and management system to resolve the issues related to the development and managing the network devices with a key focus on Augmented Reality (AR).

In [4], Nishino et al. proposed an AR assistance system for Campus Area Network management using marker-based AR system for detecting network devices. In case of large network system, marker-based AR system sometimes posed some kind of problem for installing distinct markers for every device on the network. But naturally, the campus Area Network was relatively small. The system had to store the information beforehand into a database for displaying the information.

Haramaki et al. [5] proposed a system that will assist network administrators using an AR platform in their HMD device. They used a marker for identifying the network devices. The visualization contents like IP addresses and VLAN configuration information were stored in a server named information management server. The system is configured beforehand by registering all network topology information in this server. They later introduced a marker-less system for reducing the load. This system provides a hands-free mobile operation environment and helps the user to concentrate on the administration tasks in a real machine room.

In [6], the authors modified the current system by using the Wi-Fi signal. The user will simultaneously sense the Wi-Fi signal from his current location and thus the system will identify the network device near the user. Both marker-based and marker-less AR have been introduced and implemented for the detection of devices. Though implementing marker-less AR requires much more complex programming, it provides better flexibility for the client for performing their tasks efficiently.

In [7], Flinton et al. proposed an AR system where we could acquire the information from the detected device. After detecting a device by a specific marker on that device it uses SSH or Telnet command to retrieve the network information of that device. That information is then displayed over the real-time image of that device.

We have observed that various researchers tried to implement the network management system differently. Each framework had its limitations and strengths. But no one provided a layered framework to t in any type of AR technologies for network management which would help the developers in implementing the system easily. The spectrum of users will also increase by this method. Another challenge of this method is maintaining an information server and connecting the server with the AR application. A dynamic update and maintenance would reduce the burden of the network administrators by a huge margin.

3 Proposed Framework

After an extensive study on the design and development of network management system architecture and framework with the help of AR, we analyzed some limitations of the prevailing frameworks. We have proposed a layered framework to mitigate the limitations of the previous frameworks and which will help in the smooth operation of AR technology [8]. Firstly we designed a framework that can fit into any problem raised or in any research area and provide a solution with AR technology.

3.1 Framework for Platform Independent Model

A platform-independent model (PIM) in software engineering is a model of a software system or business system that is independent of the specific techno-logical platform [9]. We have designed a framework for PIM in Fig. 1.

For the smooth integration of the AR technology, there are some requirements that the system and clients need to ensure. Firstly, the client needs to understand the problem statement and need to focus on the subject domain with which need to be integrated into the AR technology.

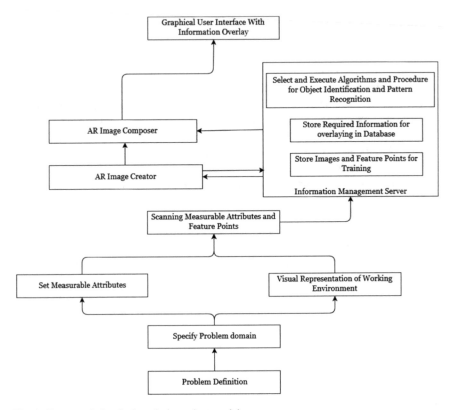

Fig. 1 Framework for platform independent model

For creating a visual representation of the working environment we need a camera to capture the video stream of the working environment. This operation can be easily done by the smartphone cameras which is readily available to most users [10].

Users then need to set various features which will be detected by the system after we have access to the visual representation of the environment. These attributes can act as the trigger or marker which will initiate the AR operation of the system [11].

Then the objects which are required to be scanned are scanned by the camera. By scanning, we imply extracting the feature points of the object. The scanning process is based on the Image registration process. The next step in this process is using the data obtained in the previous step while detecting feature points to create a real-world coordinate system. If there is partial information about the geometrical coordination in the image space then Simultaneous Localization Mapping (SLAM) algorithm is used to map relative positions. If we do not have any information about the scene geometry then the SLAM algorithm will not work. In that case, we have to undertake the Structure from Motion (SFM) method like Bundle Adjustment. These complicated processes are done by various AR SDK which performs the back-end calculation and execution of these algorithms to present us an object with detected feature points [12].

After scanning the object we need to store the feature point information in a database or server. There are many methods for executing this process. The most common method is using the database of SDKs which have the service of saving the targets or objects. But this internal database saves only the object information and sometimes that is not sufficient. Because some applications may need to access information from a different database to access the information for displaying on the virtual display. So this integration is done using PHP and C#code in the engine where we are performing the application [13].

The next step in this framework is related to augmenting the real world image with virtual information. Once the application recognizes the required object accessing the information from the database it starts the execution of the AR application as per the requirement. It starts accessing the information required to display from the database and places them in the image space.

After creating the augmented image it is rendered in the engine and we then have a view of virtual information over live video streaming of the working environment. This rendering may take some time but excessive information will generate higher response time which will diminish the usefulness of the application.

The system administrator needs to develop a user-friendly GUI to display the information. There should not be an overload of information which will create confusion among the user. The information should be precise such that the user can easily understand the information the application trying to convey.

3.2   Framework for Platform Specific Model

For developing an AR app for Network Device Management we introduced a platform-specific layered framework. We divided our framework into four layers: User Layer, Analytical Layer, Cloud Layer, Augmentation Layer. Each of the layers here has different functions. We have differentiated each of the tasks in different layers for the ease of understanding and visualization. Figure 2 represents our proposed platform specific model.

The user layer is responsible for executing the local and user side operations. Multiple AR devices such as mobile, tablets or Head Mounted Display (HMD), simultaneously start with sensing the real environment, producing raw videos, and capturing user's gestures via their cameras and sensors. The video streams contain the raw video data which is further duplicated into two copies. One is stored in the up-link cache for data transmission and the other one is stored in the local cache for subsequent processing. This raw video data will be sent to the analytical layer.

The analytical layer plays a critical role in this framework. As it receives the uploaded data from AR devices, the raw videos are then converted into several frames through an image processing module. The clipping module slices one representative frame (or image) from each raw video for subsequent processing. Features of various

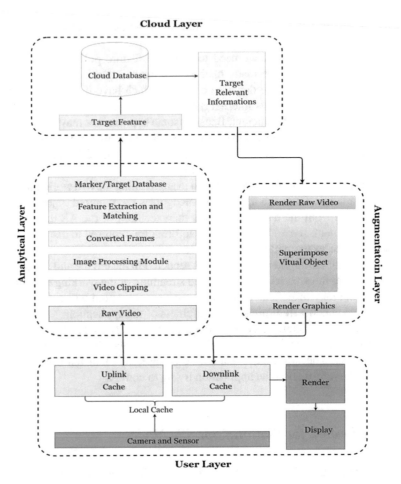

Fig. 2 Layered representation of platform specific model

interests are being extracted from these frames. Once an image is fed into the feature extraction and matching module of the analytical layer, the feature extraction algorithm will immediately search its inherent salient interest points, which are used to estimate the similarity between this image and the standard pre-stored images in the databases. After obtaining the best-matched standard image, the result will be sent on the cloud layer for further processing.

The cloud layer contains a large cloud database for storing the additional data that are not cached in the analytical layer due to its limited memory size. It contains all the information about the network devices along with the information of the target/marker. Based on the information given by the analytical layer the cloud layer search for necessary and relevant data of the image and pass that information in the augmentation layer for the final processing.

The augmentation layer provides the user with the final augmented model. It collects necessary information i.e. 3D model les, model-related data, etc. from the cloud database. After obtaining this information this layer superimposes the augmented object in the real-time raw video with the required information and sends it to the down-link cache in the user layer. The down-link cache renders the video and shows the output in the display to the end-user.

4 Proposed Application Prototype

In this section, for developing an application we have shown the basic workflow at run-time of our proposed application model in Fig. 3.

In Fig. 3, a camera-equipped device (left top) reads a video stream that is rendered as a video background to generate a see-through effect on the display. The camera device will be placed in front of the network device which the user wants to identify. In the proposed model we used Vuforia as a tracking module because it incorporates computer vision technology to recognize and track planar images and 3D objects in real-time which is fast and accurate [14]. The tracker can load and activate multiple data-set at the same time which contains the computer vision algorithms that detect and track real-world objects in-camera video frames. Device Database stores marker targets in the device itself and the cloud database stores the target in the cloud. After tracking and detecting the required object is done, data are sent into a classifier that is used for the learning process to classify new records (data) by giving them the best target attribute (prediction).

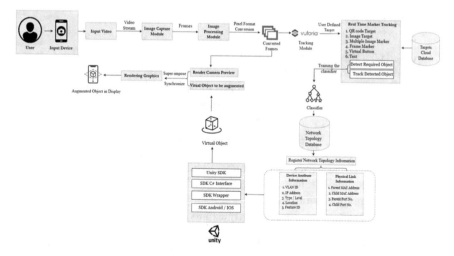

Fig. 3 Proposed application model

Fig. 4 Physical information

The user has to provide all the information needed to be displayed in the Network Topology Database before this operation. The recorded data fall into two types: physical topology information and logical topology information. The physical information includes the individual information of each network device such as MAC address, machine name, manufacturer name, model number, and device type (router, L3 switch or L2 switch). The Device Attribute Information includes IP address, VLAN configuration, location, feature ID assigned to the registered devices.

For augmenting the virtual object we used Unity 3D which is a powerful cross-platform 3D engine and provides custom resources for building AR applications. After acquiring all the information we superimpose the virtual object created in Unity with the raw input and after rendering graphics it shows the output in the display. We used C#language while creating virtual objects in Unity.

4.1 Developed Prototype

We worked on a Proof-of-Principle (PoP) Prototype also known as Proof of Concept (PoC) prototype for establishing the foundation of our conceptual frame-work. This prototype contains some key functionalities of the intended design, but it does not contain all the functionalities of the final product.

Marker-less Prototype We scanned a router as our network device in this process. For scanning the router at first we used an android application developed by Vuforia named Vuforia Object Scanner. The object was scanned under moderately bright and di use lighting avoiding any direct lighting. Because scanning objects with reflective surfaces under direct lighting can introduce areas with no tracking points. During the scanning session, the Object scanning target defines our object target's position and orientation relative to the origin of its local coordinate space. A printable object scanning target is included in the Vuforia Object Scanner download package. Thus we have scanned our required object and upload the required *.od le in the Vuforia database. This is the process of detecting the object for the implementation of Markerless AR using the Vuforia platform. Figures 4 and 5 demonstrate the outcome of our developed prototype.

Fig. 5 Port identification

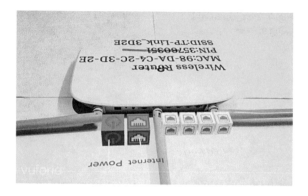

Fig. 6 Marker image

Marker-based Prototype For developing a Marker Based AR we used the image in Fig. 6 as a marker for identifying our device and attached it to our object. Vuforia has a target database where we can upload our marker which can later be detected while capturing the live stream of video of the working environment. First, we uploaded our image to the target database. After the integration of Unity 3D and Vuforia, we downloaded the marker image which was uploaded in the target database in the Unity editor. In the Unity editor, we designed the text, cubes, and lines which will be required to show the different ports of the router. Then, we augmented the router's different ports concerning that marker and placed the designed virtual information in desired places of the screen. After that, we built an application using Unity 3D and C#in Visual Studio code that runs on the smartphone and displays our required virtual information. Using this app, we could recognize the router's different ports. Figure 7 shows the prototype of the marker-based approach.

5 Conclusion

Our objective was to build a framework that can act as a guide for developing a Network Maintenance system based on AR. We proposed frameworks for both

Fig. 7 Port identification
using marker

platform-independent model and platform-specific model and showed the implementation concept of developing systems following that framework. The developed prototype was a Proof of Concept (PoC) for our proposed framework. The main purpose of our prototype was to show that following the steps of our framework we could build a system for network management with AR technologies. This technology eliminates the needs of a maintenance operator for assisting others in network issues. It can provide the detail information of the network devices to any user so that they can alone handle situations may arise in data centers.

Our research contains some limitations such as the proposed framework was not implemented in a real distributed network as well as the system was not tested by the technical persons. Therefore the friendliness of the system to users was not measured. This also warrants a real-time simulation. We will further improve our system by integrating a database with our developed prototype. We can also use the system for detecting any error in network connection and indicate the point of error which will reduce the time required for error detection by the network administrator. The frameworks will act as the basis of further improvement in incorporating more AR functionalities for network management.

References

1. Augmented Reality—Introduction and its Real World Uses, 3 Pillar Global, 2020. [Online]. Available https://www.3pillarglobal.com/insights/augmented-reality-introduction-and-its-real-world-uses. Accessed 18 Feb 2020
2. Augmented reality, En.wikipedia.org, 2020. [Online]. Available https://en.wikipedia.org/wiki/Augmented_reality. Accessed 18 Feb 2020
3. What Is Network Maintenance?|Network Maintenance Plans & Tips, Worldwide Services, 2020. [Online]. Available https://worldwideservices.net/network-maintenance-guide-upkeep/. Accessed 18 Feb 2020
4. Nishino H, Nagatomo Y, Kagawa T, Haramaki T (2014) A mobile AR assistant for campus area network management. In: 2014 Eighth International Conference on Complex, Intelligent and Software Intensive Systems, pp. 643. 648, IEEE
5. Haramaki T, Nishino H (2015) A network topology visualization system based on mobile AR technology. In 2015 IEEE 29th International Conference on Advanced Information Networking and Applications, pp. 442. 447, IEEE

6. Haramaki T, Nishino H (2015) A device identification method for AR-based net-work topology visualization. In 2015 10th International Conference on Broadband and Wireless Computing, Communication and Applications (BWCCA), pp. 255. 262, IEEE

7. Flinton C, Anderson P, Shum HP, Ho ES (2018) NETIVAR: NETwork information visualization based on augmented reality. In: 2018 12th International Conference on Software, Knowledge, Information Management & Applications (SKIMA), pp. 1. 8, IEEE

8. Bauer M, Bruegge B, Klinker G, MacWilliams A, Reicher T, Riss S, Sandor C, Wagner M (2001) Design of a component-based augmented reality framework. In: IEEE and ACM International Symposium on Augmented Reality, pp. 45. 54, IEEE

9. Platform-independent model, En.wikipedia.org, 2020. [Online]. Available https://en.wikipe dia.org/wiki/Platform-independent_model. Accessed 18 Feb 2020

10. Kasetty sudarshan, Sneha (2017) Augmented Reality in Mobile Devices

11. Hirzer M (2008) Marker detection for augmented reality applications. In: Semi-nar/Project Image Analysis Graz, pp. 1. 2

12. Silva RL, Rodrigues PS, Mazala D, Giraldi G (2004) Applying object recog-nition and tracking to augmented reality for information visualization. Technical Report. Technical report, LNCC, Brazil

13. Reitmayr G, Schmalstieg D (2003) Data management strategies for mobile aug-mented reality. Citeseer

14. Amin, Dhiraj Govilkar, Sharvari (2015) Comparative study of augmented reality Sdk's. Int J Comput Sci Applic 5:11–26. 10.5121/ijcsa.2015.5102

A Safety Helmet Detection Method Based on the Combination of SSD and HSV Color Space

Wei Wang, Song Gao, Renjie Song, and Ziming Wang

Abstract Wearing safety helmet is a protective measure, operators must wear them when they entering the construction area. Safety helmet can significantly reduce the probability of brain injury of workers, it ensures the safety of operators to some extent. Workers do not wear helmets in some actual construction processes, so it is necessary to detect the helmets in the construction area. Only relying on manual identification not only consumes a lot of manpower and material resources, but also is inefficient. So manual identification cannot meet the requirements of real-time monitoring. In this paper, a method of helmet detection based on human body recognition is proposed in which SSD neural network model is trained to recognize operators quickly. According to the geometric characteristics of human body and combined with HSV color space and morphological processing, helmet wearing is detected. Through the monitoring camera to obtain the video stream of actual work environment as the experimental data, the effectiveness of the above method is verified by the experiment.

Keywords Helmet detection · SSD · HSV color space · Deep learning

1 Introduction

Safety is the eternal theme of engineering operation. However, due to the weak safety awareness of operators, inadequate supervision of safety supervisors and other reasons, accidents occur every year due to not wearing safety helmet.

In recent years, researchers have applied sensor and image processing technology to real-time identification and tracking of people, materials and machines in construction site. Gong Rui et al. [1] used the satellite positioning system to locate the safety

W. Wang · S. Gao
State Grid Jilin Electric Power Co. Ltd, Electric Power Research Institute, Changchun, China

R. Song · Z. Wang (✉)
School of Computer Science, Northeast Electric Power University, Jilin, China
e-mail: wziming_96@163.com

helmet position, but the hardware equipment was too expensive to be widely used. Zhu Fuyun et al. [2] used the HOG feature to identify pedestrians, and used the template matching method to detect safety helmets. However, the calculation of HOG feature is too large to meet the needs of real-time monitoring, and the recognition effect of blocked pedestrians is poor, resulting in missed detection. Pang Xia et al. [3] first used skin color and eye and mouth position to locate the face, combined with YCbCr color space to detect the helmet, but in practical application, due to the different positions of operators, it is difficult to display the complete face in the image, resulting in positioning failure, which seriously affects the detection effect.

Based on the existing research results and practical application, this paper proposes a safety helmet detection technology combining SSD [4] operators detection and HSV color model. This technology combines the advantages of high accuracy and speed of SSD, small influence of light on HSV color model, and can accurately and quickly detect the operators in the scene. Combined with the geometric characteristics of human body and HSV color space, morphological processing is carried out to monitor the operators wearing safety helmet in real time, so as to avoid the occurrence of wrong inspection of operators holding safety helmet.

2   Implementation of Safety Helmet Detection Algorithm

Because there is no mature data set for helmet wearing detection, it is necessary to collect and preprocess data by ourselves. The SSD model is trained by the processed data, which is used to detect operators and get the head position according to the geometric characteristics of human body. Because the color of safety helmet is fixed, the independence of three channels in HSV color space is used to detect. Finally, the condition of wearing safety helmet is detected by morphological processing. This part describes the detailed implementation process of the algorithm proposed in this paper, and the algorithm flow chart is shown in Fig. 1.

2.1   Experimental Data Acquisition and Preprocessing

Data preprocessing is very important in the construction of network model, and often determines the training results. Of course, for different data sets, preprocessing methods will have more or less particularity and limitations. As one of the benchmark data, Pascal voc2012 is frequently used in object detection, image segmentation network contrast experiment and model effect evaluation.

The video stream is recorded by the DS Series surveillance camera of Hikvision, with the pixel of 640×480. After that, the video stream is saved as pictures frame by frame. The data set is annotated by label Image, and divided into data set and training set. The XML file is generated. Finally, the data set of VOC2012 format is obtained.

Fig. 1 Algorithm flow chart

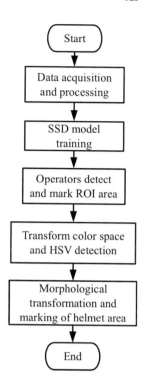

2.2 SSD Operator Detection

SSD introduces a single-stage detector, which is more accurate and faster than the previous algorithm Yolo. Instead of using RPN and pooling operations, it uses a small convolutional filter to apply in different feature map layers to predict the category of bounding box. More importantly, it can get better detection results in smaller input pictures. Test results on multiple datasets (PASCAL, VOC, COCO, ILSVRC) show that SSD can get higher mAP value. The SSD network architecture is shown in Fig. 2.

In this paper, SSD is selected as the network model of detection workers, which has obvious speed advantage compared with Faster R-CNN. The data set is imported

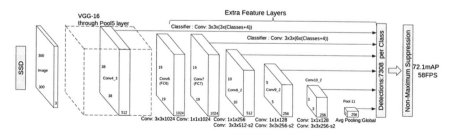

Fig. 2 SSD network architecture

into TensorFlow, a deep learning framework, for SSD model training, and finally the SSD detection model meeting the requirements of small loss value is obtained.

2.3 HSV Helmet Detection

RGB is the most common color model. It uses the linear combination of three colors components to represent color. Any color is related to these three components. However, these three components are highly correlated. To adjust the color of an image, you must change three components. Images acquired in natural environment are easily affected by natural light, occlusion and shadow, that is, they are sensitive to brightness. The three components of RGB color space are closely related to brightness, that is, as long as the brightness changes, the three components will change accordingly, without a more intuitive way to express. Therefore, RGB color space is suitable for display system, but not for image processing.

Because RGB channel cannot reflect the specific color information of the object very well, and compared with RGB space, HSV space can express the brightness, hue and brightness of the color intuitively, which is convenient for color contrast. The parameters of color in this model are hue (H), saturation (S), brightness (V). Because each color channel of HSV is independent, and the influence of light and shadow on H and S components is very small, it can eliminate the influence of weather and other reasons to a certain extent in practical application, so as to better identify the safety helmet with color characteristics. The HSV color model is shown in Fig. 3.

The conversion formula (1), (2) and (3) give the conversion formulas from r, g, b values of RGB color space to h, s, v values of HSV color space.

$$\begin{cases} Max = \max\{r, g, b\} = v \\ Min = \min\{r, g, b\} \end{cases} \tag{1}$$

Fig. 3 HSV color space model

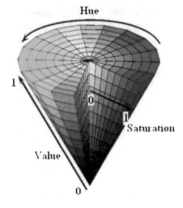

$$h = \begin{cases} 0°, if\,max = min \\ 60° \times \frac{g-b}{max-min} + 0°, if\,max = r\,and\,g \geq b \\ 60° \times \frac{g-b}{max-min} + 360°, if\,max = r\,and\,g < b \\ 60° \times \frac{b-r}{max-min} + 120°, if\,max = g \\ 60° \times \frac{r-g}{max-min} + 240°, if\,max = b \end{cases} \quad (2)$$

$$s = \begin{cases} 0, if\,max = 0 \\ \frac{max-min}{max} = 1 - \frac{min}{max}, otherwise \end{cases} \quad (3)$$

Through the above conversion process, the result of converting RGB image to HSV image is shown in Fig. 4.

In the process of helmet detection, if color space conversion is performed on the whole image, it cannot meet the needs of real-time monitoring algorithm. Therefore, ROI area [6] is added to the algorithm, that is, only the upper third of the area where the construction personnel are detected by SSD are used for color space conversion calculation, which greatly reduces the conversion area and improves the operation speed of the algorithm. After recognition, it is represented by a binary image, but there will be noise in the ROI region that meets the specified HSV region.

Open operation is mainly used to keep a certain structure operation, which can eliminate noise and small connected area without changing the main body of the image. This algorithm uses morphological open operation to eliminate noise. The algorithm results are shown in Figs. 5 and 6.

Fig. 4 RGB and HSV color space image diagram

Fig. 5 Results of wearing safety helmet

Fig. 6 Results of not
wearing safety helmet

The red box in Fig. 5 indicates the location of the safety helmet, the green box indicates the operator wearing the safety helmet, and the blue box in Fig. 6 indicates the operator not wearing the safety helmet. From the recognition results, it can be seen that the algorithm can accurately identify the situation that the operator wears the safety helmet in real-time, and it also has better recognition for the smaller object in the image.

3 Conclusion

In this paper, the SSD neural network model is used to identify the constructors in the video stream, which not only reduces the error detection rate of the safety helmet, but also demarcates the ROI of the safety helmet detection combined with the geometric characteristics of the human body, reduces the calculation amount brought by the color space conversion, and improves the detection speed of the algorithm. By using the independent advantages of HSV three channels, it reduces the influence of light, shadow and other factors on the recognition results in the practical application environment and the noise of false detection is eliminated and the small connected area is separated by using morphological operation. The experimental results show that the algorithm is fast, can meet the requirements of real-time monitoring, and has good robustness.

References

1. Gong Rui, Lu Jizhao (2019) Application of Beidou high precision positioning technology in electric power patrol inspection—intelligent safety helmet. Inf Technol Inf 2019(10):64–66+70
2. Fuyun Z, Xianglong D et al (2018) Research on algorithm of helmet detection in substation based on video monitoring. Power Grid Clean Energy 34(12):71–76
3. Pang Xia, Wang Lirong (2017) Helmet recognition method based on color and shape. 2017 Smart Grid Inf Construc Sympos 2017:50–51+54
4. Liu W, Anguelov D, Erhan D et al (2016) SSD: single shot multibox detector. European conference on computer vision, 21–37

5. Fu Changfei, ye bin et al (2020) Moving object recognition based on HSV color space. Control and information technology, 1–5
6. Bei H, Li Y et al (2017) A fast image enhancement-based method for bottle pouring detection in production line. J Beijing Uni Inf Technol (Natural Science Edition) 32(5):39–44

Vision-Based Diagnosis and Location of Insulator Self-Explosion Defects

Wei Yang, Zhiwei Gu, Renjie Song, and Yingjie Li

Abstract Aiming at the insulator image of transmission line acquired by UAV, a vision-based insulator self-explosion defect detection and location method was proposed. First, superpixel segmentation is performed on the insulator image based on local texture features, and the saliency map of the insulator string is obtained by using the color feature different saliency and multi-scale optimization. Then, the salient image is binarized and morphologically processed to obtain a binary image. Finally, vertical projection Method to identify and identify the location of insulator defects. The experimental results show that the method can accurately identify the fault point of insulator strings. By comparing with two commonly used insulator self-explosion fault detection methods, the validity and reliability of the proposed method are proved.

Keywords Superpixel · Saliency · Multiscale · Vertical projection · Defect diagnosis

1 Introduction

As a special insulation control, insulators play an important role in the safe and stable operation of transmission lines [1]. Due to the long-term exposure of the insulator to the field and the influence of the natural environment and heavy mechanical loads, the insulator is extremely prone to failure problems such as self-detonation and damage [2]. Once the insulator fails, it will cause hidden danger to the operational safety of the transmission line. Therefore, regular monitoring of insulator conditions and timely detection of insulator failure are of great significance to the safe operation of power systems.

W. Yang · Z. Gu
Quzhou State Grid Power Supply Company, Quzhou 324000, China

R. Song · Y. Li (✉)
School of Computer Science, Northeast Electric Power University, Jilin 132012, China
e-mail: lyj_neepu@163.com

Nowadays insulator fault detection methods are mainly divided into 4 categories: Through manual inspection, using a three-dimensional laser rangefinder, using ultraviolet equipment and based on computer vision [3]. In the current research of intelligent power grid, computer vision technology has been widely used in the field of intelligent inspection and online monitoring of power equipment. The main research ideas in the field of insulator fault diagnosis and identification are basically divided into 2 steps: ① Locate the insulator string in the aerial image; ② Recognize the insulator string. Literature [4] locates insulator strings through the idea of weakly supervised fine-grained classification, and combines the MFIFIN network full-label information training model to identify the insulator strings, but the current network training is more complicated and cannot be trained end-to-end. Literature [5] uses the YOLOv2 deep learning network to identify insulator strings, and uses related image processing algorithms to achieve fault diagnosis of the insulator's self-detonation position. A variety of methods have been used to identify fault points, but the network training time is long and the number of iterations is high. YOLOv2 The network has weak ability to identify dense targets. In literature [6], the positioning of insulator strings is completed through the visual geometric group network model, and the ResNet model is used as a classifier to diagnose the insulator strings. Since there is no public insulator string data set, training the network is time-consuming and can cause overfitting problems because the data set is too small.

This paper proposes a vision-based method for detecting and locating insulator self-detonation defects based on the self-detonation fault of insulators. The local texture feature saliency area detection algorithm is used to obtain the insulator candidate area. The vertical projection is used to determine whether the candidate area is an insulator string. The relationship between the wave crests, to realize the accurate identification of insulator self-detonation.

2 Extraction of Candidate Region of Insulator String

2.1 Superpixel Segmentation Based on Local Texture Features

The local operator can well describe the local texture features of the image, and has the characteristics of simple calculation and strong expression of the image texture features. In this paper, the SLIC segmentation algorithm based on texture features, color and space is used to segment the image, and then used for saliency detection.

First convert the image from RGB image to LAB color space, initialize the clustering center, and calculate the similarity measure D of the two pixels combined with the texture features:

$$D = d_c + \frac{m}{s}d_s + cd_w$$

Fig. 1 Comparison of original image and superpixel segmentation

In the formula, the arithmetic formula of d_c, d_s, d_w is:

$$d_c = \sqrt{(l_i - l_j)^2 + (a_i - a_j)^2 + (b_i - b_j)^2}$$

$$d_s = \sqrt{(x_i - x_j)^2 + (y_i - y_j)^2}$$

$$d_w = \sqrt{(w_i - w_j)^2}$$

In the formula, d_c is the color distance of two pixels in Lab space, d_s is the space distance, and d_w is the differential distance of local texture features. S, m, and c are constants, which comprehensively represent the weight of space, color, and texture features in the similarity measure.

In addition, vl, a, bv, (x, y), w respectively represent pixel color features, pixel space coordinates and pixel texture features.

For the above algorithm, when the number of superpixels is set to $K = 150$ at the same scale, an example of the results of processing two original images containing insulator strings is shown in Fig. 1. It can be seen from the image results that the superpixel image obtained by the processing has the characteristics of high consistency, uniform characteristics, and can better reflect the pixels inside the superpixel area.

3 Identify Candidate Regions for Insulator Strings

Different weights are assigned according to the differences in background superpixel features. $B_p(r) = 1 - \exp(-B_{con}^2(r))$, where $B_p(r)$ is the weight of the background superpixel, which represents the strength of the background attribute. Combine the weight value to calculate the saliency of the superpixel $C_t(r_i)$ based on the background prior.

Fig. 2 Binarized
morphological processing

The number K of superpixels is set to three different scales, 150, 300 and 600, respectively, so that while retaining the overall structure information, it can retain more local details in the saliency detection. Multiscale fusion of the saliency of the above three superpixels at different scales $RC(k)$:

$$RC(k) = \frac{1}{n} \sum_{i=1}^{n} RC_t^i(k)$$

Among them, $RC_t^i(k)$ is the significance of pixel k at the i th scale. Then determine the superpixel saliency map by normalizing $RC(k)$, and finally use bilateral filtering to denoise to get the final saliency map $S(K)$.

After obtaining the $S(K)$ saliency map of the insulator image, the adaptive threshold method is used to perform binary segmentation on the image, combined with morphological operations to remove noise in the image. Calculate the connected domain in the figure as a candidate area and mark it to prepare for the next insulator failure detection. The resulting candidate area is shown in Fig. 2.

4 Insulator Recognition and Fault Detection

Insulator strings are composed of insulator pieces of the same shape and color and the central axis are vertically arranged at equal intervals; the appearance structure can be viewed as a row of fixed length and width ellipses. The sizes vary and there are no rules at all [7]. Based on the morphological characteristics of the insulator, it is determined whether the candidate region is an insulator string. The implementation steps are as follows:

(1) First, use random uniform sampling to fit the straight line of the main axis of the candidate area, and by rotating the main axis of the candidate area to the horizontal direction, accumulate 255 pixels in the insulator candidate area in the vertical direction:

$$S(x) = \sum_{i=1}^{r} (I_2(i, x)/255)$$

In the formula, $S(x)$ represents the accumulated value of the non-zero pixels of the insulator mask in each column direction.

(2) Set the variance threshold $D_{th} = 10$ and calculate the peak value variance D^2:

$$D^2 = \frac{\sum_{i=0}^{n} (S(x_i) - avgs)^2}{n - 1}, \quad x_i \in x_{point}$$

The decision formula of the candidate region in the image is:

$$\begin{cases} D^2 > D_{th}, & \text{Candidate regions are insulator strings} \\ D^2 > D_{th}, & \text{candidate regions are non} - \text{insulator strings} \end{cases}$$

(3) Detection of dropout area. Considering the morphological wave structure of the insulator, when the candidate region is an insulator string, after the image is vertically projected, the mutation between the two peak segments is found according to certain criteria as the position where the insulator appears off the string. From the peak point sequence $(x_1, x_2, ...x_n)$, calculate the proportion of pixels with a pixel value of 255 in the insulator sub-mask between the peak segments, and find the drop-out interval as a feature:

$$g(x) = \frac{s(x_p) + s(x_{p+1})}{(x_{p+1} - x_p) * row(I_2)}$$

Based on the small $g(x)$ value at the dropped string, the algorithm for detecting the dropped string position of the insulator is constructed as follows:
$area = (\max(g(x) + \min(g(x)))/2,$
$g(x)/area < 0.5,$
There is a drop at the peak $[x_p, x_{p+1}]$. The position of the insulator detected in the graph coordinate system can be mapped to the input image inversely according to the change relationship between the image coordinates, and the final insulator defect positioning is completed.

5 Experimental Results and Analysis

In order to verify the effectiveness of the method in this paper, the algorithm in this paper is compared with the existing insulator string self-detonation recognition algorithm, and the performance of the experimental algorithm is evaluated using F-score

Table 1 Performance comparison of defect detection methods

Recognition methods	Acc/%	F_1
Literature [7]	90.6	0.689
Literature [5]	87.3	0.712
This article method	91.7	0.823

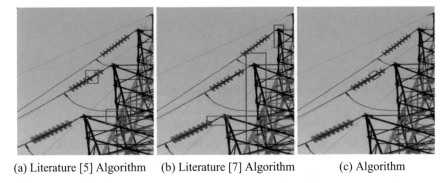

(a) Literature [5] Algorithm   (b) Literature [7] Algorithm   (c) Algorithm

Fig. 3 Comparison of insulator self-explosion defect recognition algorithms

and accuracy in computer vision evaluation indicators. The performance comparison of the algorithm and the results of the insulator self-detonation defect identification algorithm are shown in Table 1 and Fig. 3.

It can be seen from Table 1 and Fig. 3 that the method of [5] has better recognition accuracy, but there are more false detections and the comprehensive evaluation is not high; although the method of [7] has poor positioning accuracy, the comprehensive evaluation is high In literature [5]; the accuracy performance of the method in this paper is more than 91.7%, and the recognition effect is also better than the two methods in the literature.

6 Conclusion

This paper presents a vision-based method for detecting and locating insulator self-explosive defects. It considers superpixels as the basic unit, and extracts candidate regions of insulator strings based on multi-scale optimization of images and differences in color features. Through the binarization and morphological processing to obtain the binary image of the insulator region, analyze the relationship between the peaks of the projection curve to accurately locate the defect position of the insulator. During the experiment, the self-collected insulator string image was invoked as the experimental sample, and the effectiveness experiments were compared with the two insulator fault detection algorithms. Experimental results show that the method in this paper has higher recognition accuracy than the two methods, and the recognition evaluation standard achieved by the method is also higher than the two methods compared.

References

1. Zhining L (2020) Overview of common fault analysis and detection methods for transmission lines. Autom Instrum 35(01):161–164
2. Zhen B, Liu Z (2014) The recognition and localization of insulators adopting SURF and IFS based on correlation coefficient. Optik 125(20):6049–6052
3. You C (2019) Insulator fault detection method based on catenary imaging technology. Mod Urban Rail Transit 156(10):5–10
4. Pan Z, Zhang X, Yang Y et al (2020) Application of weakly supervised fine-grained classification in insulator fault identification. J Shanxi Univ 18(01):86–95 (Natural Science Edition)
5. Zhang G, Liu Z, Han Ye (2016) Automatic recognition for catenary insulators of high-speed railway based on contourlet transform and Chan-Vese model. Optik 127(01):215–221
6. da Silva PR, Gabbar HA, Junior PV et al (2018) A new methodology for multiple incipient fault diagnosis in transmission lines using QTA and Naïve Bayes classifier. IntJ Electr Power Energy Syst 103(12), 159–165
7. Zhai Y, Wang Di, Zhao Z et al (2019) Insulator string localization method based on spatial morphology consistency feature [J]. Trans China Electr Eng 37(05):1568–1578

Part IV
Internet of Things

A Decentralized Access Control Model for IoT with DID

Euihyun Jung

Abstract IoT has put a great impact on modern society and every industry, but there are still several technical issues to be resolved in realizing the vision of IoT. From these issues, Access Control (AC) is essential to make an IoT service successful because, in the IoT environment, people with different privileges will access the devices from multiple manufacturers under diverse situations. However, the existing centralized AC models have the underlying limitations from the centralized approach such as a single point of failure. The research proposes a decentralized access control model for IoT with Decentralized ID (DID) and explains how the proposed model authorizes access rights successfully without a centralized authority.

Keywords Decentralized identifier · Access control · Internet of things (IoT)

1 Introduction

Internet of Things (IoT) has influenced every industry including manufacturing, healthcare, transportation, agriculture, etc., and its adoption is increasing to become the main drive force of economy [1]. However, it is also true that there are still obstacles to keep from realizing the vision of IoT such as interoperability, the orchestration of heterogeneous things, security, privacy and so on [2]. From the obstacles, Access Control (AC) of people and things is one of the main obstacles preventing the widespread applicability of IoT [3]. Since, in IoT environment, the devices from multiple manufacturers are accessed by people who have different privileges under diverse situations, an AC ruling over an IoT service consistently is needed for the success of the IoT service.

Usually, AC defines who (i.e., subject) can use which access rights (i.e., actions) on which resource (i.e., object) [4]. To fulfill this purpose, AC has policies that describe the conditions that subject satisfy to get access rights. AC is already essential in IT domain, but its importance becomes greater in IoT environment because the situation

E. Jung (✉)
Department of Convergence Software, Anyang University, Anyang City, South Korea
e-mail: jung@anyang.ac.kr

© The Author(s), under exclusive license to Springer Nature Singapore Pte Ltd. 2021   141
H. Kim et al. (eds.), *IT Convergence and Security*, Lecture Notes
in Electrical Engineering 712, https://doi.org/10.1007/978-981-15-9354-3_14

and the role of participants in IoT are much more complex than the ordinary IT domain [5]. There have already been some AC models for IoT such as Discretionary Access Control (DAC), Mandatory Access Control (MAC), Role-based Access Control (RBAC) and Attribute-based Access Control (ABAC).

These models have their own pros and cons, but they are common in requiring a special kind of a centralized Certified Authority (CA) [6]. That means the participants in IoT environment rely so much on the centralized authority to perform a successful AC. This dependency causes several issues: scalability, managing heterogeneity, resolving object identities, managing personal data, and so on [7].

In order to resolve this issue, the research proposes a decentralized AC model using Decentralized Identifiers (DIDs) [8]. In the proposed model, every participant of an IoT application has its own identifier based on DID and proves his/her/its identity to get access rights in a decentralized way without the dedicated CA. Currently, the most commonly employed approach for authentication is Public Key Infrastructure (PKI) based on CA system which adopts a centralized trust infrastructure. Although PKI can provide participants with the secure communication channel and non-repudiation, the centralized PKIs such as the CA-based system has its own problems and limitations because it wholly relies on a central trusted party. To resolve this issue, the proposed model adopts Decentralized Public Key Infrastructure (DPKI) [9]. Instead of using a central CA, entire authority information is stored in the DID document corresponding to its owner's DID and each participant performs PKI operation in a decentralized way.

The rest of this paper is organized as follows. Section 2 describes the overview of DID, the schemas of DID document, and the authorization process which checks a peer's identity without a central CA. In Sect. 3, the evaluations of the proposed model are discussed, and Sect. 4 concludes the paper.

2   Design of the System

2.1   An Overview of Decentralized Identifier

Decentralized Identifiers (DIDs) provide the identification of humans, devices, and digital beings in verifiable and decentralized ways [8]. Due to its great potential usability, W3C has made a lot of effort to publish the standard of DID and other groups have announced their own DID schemes. In the W3C standard, a DID is similar to a Universally Unique Identifier (UUID), but it can be resolved to a standard resource named DID document describing the subject. Unlike other identification schemes, a DID document usually contains cryptographic information and the document can be stored and resolved with distributed ledgers such as Ethereum.

A DID itself is a simple text string consisting of three parts [8] as shown in Fig. 1. The first part is a URL scheme identifier fixed as "did". The second part is an identifier for the DID method which defines how a specific DID scheme can be implemented

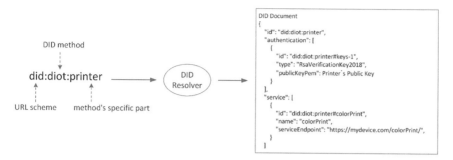

Fig. 1 The structure of a DID and the example DID document

and resolved to DID documents. The last part is a DID method-specific part that conveys the information for the resolution. Since the DID method-specific part is not restricted by the standard, an architect of a DID identification system can pose any kind of identification elements for the design purpose.

The resolution result of a DID is a DID document that describes the subject identified by the DID. The document usually contains service endpoints and a public key. These two elements are very important to achieve decentralized access control. First, the service endpoints can be extended to contain any information, therefore it specifies the policies that grant to access IoT resources in detail. Second, the public key is used to perform PKI operations for verifying and encrypting information.

2.2   Schemas of DID Documents

There are two kinds of DID documents in the proposed model. The first one is a user DID document and the second one is a device DID document.

Schema of DID Document for Users. In the schema of a user DID document, a user's DID and the user's public key are specified. In the authorization process, these elements are essential to verify the user's identity and the fidelity of the user's messages. Figure 2 shows the abbreviated user DID documents.

Schema of DID Document for Devices. In the proposed model, a device has a DID document that specifies the access rights of the services based on membership policy. For this, there are two main descriptions the DID document: the membership and the services. In the membership description, each user is put under one of the specific memberships: owner, friend, or guest. Each membership maintains the list of users' DIDs, but this list does not force or specify the access rights of the device. The actual descriptions for the access rights are written in the parts of services that the device provides.

Each description of services specifies a name, a DID for the service, a service endpoint URL, and permission. The DID is used to indicate the service itself and the service endpoint URL is used for the outside service caller. The permission

```
{
    "id": "did:diot:john",
    "authentication": [
            {
                "id": "did:diot:john#keys-1",
                "type": "RsaVerificationKey2018",
                "controller": "did:diot:oneaboveall",
                "publicKeyPem": John's Public Key
            }
    ]
}
```

Fig. 2 The example of the abbreviated user DID document

specifies which member can use the service. For example, since a user, "John", is in the "friend" membership, "John" can use "mono-printing". However, "John" cannot use "color-printing" because "color printing" is only allowed to users who get the "owner" membership (see Fig. 3).

```
{
    "id": "did:diot:printer",
    "authentication": [
            {
                "id": "did:diot:printer#keys-1",
                "type": "RsaVerificationKey2018",
                "controller": "did:diot:oneaboveall",
                "publicKeyPem": Printer's Public Key
            }
    ],
    "service": [
            {
                "id": "did:diot:printer#membership",
                "serviceEndpoint": "https://diot.io/membership/",
                "owner": ["did:diot:mary"],
                "friend": ["did:diot:john", "did:diot:tom"],
                "guest": []
            },
            {
                "id": "did:diot:printer#colorPrint",
                "name": "colorPrint",
                "serviceEndpoint": "https://mydevice.com/colorPrint/",
                "permission": ["owner"]
            },
            {
                "id": "did:diot:printer#monoPrint",
                "name": "monoPrint",
                "serviceEndpoint": "https://mydevice.com/monoPrint/",
                "permission": [ "owner", "friend"]
            }
    ]
}
```

Fig. 3 The example of the abbreviated device DID document

2.3   A Process of Authorization

The authorization process is performed with DID and DPKI as shown in Fig. 4. In the process, when a user, "John" wants to use a printer, he gets the DID document from a DID resolver with the printer's DID to acquire the public key of the device (in the step (1) and (2)). The DID resolver maintains all DID documents and resolves a DID request to the corresponding DID document. After acquiring the device's public key, "John" makes a payload containing his DID and timestamp. Then, he generates the signed hash of the payload with his private key and sends the payload with the signed hash as a request (in the step (3)).

After receiving the request, the printer resolves John's DID to get John's DID document from the DID resolver and it verifies the fidelity of the request by checking the signed hash with John's public key contained in the John's DID document (in the step (4) and (5)). If the verification is successful, the printer browses the device DID document to decide which services are allowed to "John". Then, the printer makes a payload containing a list of the names and the endpoints of allowed services, a valid time, and a nonce encrypted with John's public key. After then, the printer generates the signed hash of the payload with the printer's private key and returns the payload with the signed hash as a response (in the step (6)).

When "John" gets the response, he verifies the payload with the signed hash with the printer's public key and decrypts the nonce in the payload with his private key. Then, "John" makes a token containing his DID and a hash from his DID and the nonce. After then, he generates a signed hash of the token with his private key and requests the target service with the token and the signed hash (in the step (7)).

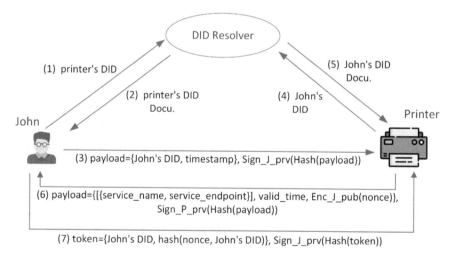

Fig. 4 The authorization process of getting the access right of the printer's services allowed to John

3 Evaluation

3.1 Trust Between Peers

In order to establish the trust between peers in a trustless environment such as IoT, the proposed model adopts the combination of "digital signature", "public key cryptography" and "hash fingerprinting". First, the model uses a "digital signature" to prove the origin of a packet. In the step (3), (6), and (7) in Fig. 4, a sender appends the signed hash with its private key and a receiver checks the validity of the signed hash with the sender's public key from the sender's DID document. With this method, a peer can prove its identity to the other side without the intervention of a centralized authority.

Second, the model uses "public key cryptography" to prevent unauthorized service requests. Since a device's DID document can be publicly accessed, someone can try to make an unauthorized service request using the service endpoints from the DID document. To prevent this kind of unauthorized request, the device generates a one-time nonce that can be used to make a token for the authorized service request in the step (6) and (7). The nonce is safely delivered to the user because the nonce is encrypted with the user's public key and only the user can decrypt the nonce.

Lastly, the model adopts "hash fingerprinting" to guarantee the fidelity of the payload. In an open IoT environment, it is possible for a hacker to eavesdrop and modify packets, so the fidelity of the message should be guaranteed. For this, the model uses a hash fingerprinting to check the fidelity of the message. If some contents are modified illegally or accidentally, the receiver can easily detect the modification by looking into the given hash fingerprinting.

3.2 A Light Burden on Management and Differentiated Membership Services

In IoT environment, there is a huge number of devices that have to assign their access rights to tremendous users. Even worse, devices may have multiple services, so the access control process becomes much more complicated. In a hypothesis IoT application where 100,000 devices having five services should provide 100,000 users, the number of records for the access control will be 50,000,000,000 in the worst case. Compared to this, the proposed model separates the users' membership from the devices' profile. This separation can dramatically reduce a manager's burden to administrate the access control information. The maximum number of DID documents will be 200,000 including devices' and users' documents.

Another advantage of the model is that it can provide differentiated membership services on the same device. Since each device can specify the different membership on users for its services, users can experience different services according to situations or locations. In addition to this, if a user wants to give temporarily an access right of some service to a visitor, the user can easily do it by upgrading the visitor's membership of the target service. This ability can be conveniently used in Smart City or Smart Home applications.

4 Conclusion

Although IoT has put a great impact on modern society and industry, there are several issues to be resolved in realizing the vision of IoT. From the issues, AC is one of the main obstacles, so several models have been announced to provide the AC for IoT. However, since the existing models are based on the centralized model, they have the same problems of a centralized approach such as scalability.

The proposed model provides the AC for IoT in a decentralized way with DID. Unlike the existing models, the model provides the AC without the intervention of any kind of a centralized CA. Therefore, this model has some advantages over the existing models. First, it can resolve the underlying problems of the centralized approach such as scalability or the single point of failure. Second, it can reduce the burden of management due to the separation of users' and devices' profile. Lastly, the differentiated membership in the model can be applied to diverse IoT scenarios.

However, since the proposed model is in its early stages, there are some points to be improved. Most of all, the model might be in danger if an unauthorized modification of DID documents. Currently, the DID resolver is responsible to manage the documents securely, but this may not a perfect solution against the hacking as with all centralized systems. Therefore, the author has a plan to adopt blockchain as a backend repository for DID documents instead of ordinary databases. Since blockchain is capable of storing and accessing data in immutable and secure ways, the adoption of blockchain will be helpful to complement the security of the proposed model.

Acknowledgements This research was supported by Basic Science Research Program through the National Research Foundation of Korea (NRF) funded by the Ministry of Education (2018R1D1A1B07049930). Icons are made by Freepik from www.flaticon.com.

References

1. Da Xu L, He W, Li S (2014) Internet of things in industries: a survey. IEEE Trans Ind Inform 10(4):2233–2243
2. Li S, Da Xu L, Zhao S (2015) The internet of things: a survey. Inform Syst Front 17(2):243–259
3. Liu J, Xiao Y, Chen CP (2012) Authentication and access control in the internet of things. In: 32nd International Conference on Distributed Computing Systems Workshops. IEEE, pp 588–592
4. Ronald F (1978) On an authorization mechanism. ACM Trans Database Syst 3(3):310–319

5. Gusmeroli S, Piccione S, Rotondi D (2013) A capability-based security approach to manage access control in the internet of things. Math Comput Model 58(5–6):1189–1205
6. Ouaddah A, Mousannif H, Elkalam AA, Ouahman AA (2017) Access control in the Internet of Things: Big challenges and new opportunities. Comput Netw 112:237–262
7. Bertin E, Hussein D, Sengul C, Frey V (2019) Access control in the Internet of Things: a survey of existing approaches and open research questions. Ann Telecommun 74(7–8):375–388
8. Drummond R, Manu S, Dave L, Christopher A, Ryan G, Markus S (2019) Decentralized identifiers (DIDs) v1.0 cord data model and syntaxes. W3C Working Draft 09 December 2019. https://www.w3.org/TR/did-core/. Last accessed 10 Feb 2020
9. Christopher A, Arthur B, Vitalik B, Jon C, Duke D, Christian L, Pavel K, Jude N, Drummond R, Markus S, Greg S, Noah T, Harlan TW (2015) Decentralized public key infrastructure. A White Paper from Rebooting the Web of Trust

Development of Wireless Sensor Nodes to Monitor Working Environment and Human Mental Conditions

Nobuyoshi Komuro, Tomoki Hashiguchi, Keita Hirai, and Makoto Ichikawa

Abstract This paper proposes and develops wireless sensor nodes to monitor working environments and human emotions estimation. The proposed system collects indoor environment data, which concern human perception, through Wireless Sensor Network. Human emotions are estimated based on the Machine Learning techniques and collected sensor data. In other words, the proposed system estimates the human emotions without image data form camera sensor or vital data from wearable sensors. Experimental results show that the proposed system achieves over 80% estimation accuracy by using multiple kinds of sensors. It is seen from the experimental results that the developed sensors are helpful to estimate human emotions.

Keywords Wireless sensor network (WSN) · Internet of things (IoT) · Human emotion estimation · Indoor environment data · Machine learning

1 Introduction

During these decades, many kinds of wireless communication techniques have been proposed and investigated [1–6]. Along with miniaturization and price reduction of wireless devices, the Internet of Thins (IoT) is getting much attention [7–18]. A lot of electric devices are connected to the Internet by realizing the idea of IoT. The IoT enables physical objects and/or space to communicate with each other. The IoT enables us to get various kinds of environment data, which can be used for big data analysis. The IoT can be applied to various kinds of applications, such as smart home, smart building, smart health care, and smart rearing [13–18].

It is expected that combining various kinds of data, which is obtained by IoT, with Machine Learning and/or big data analysis helps us to develop new services

N. Komuro (✉) · T. Hashiguchi · K. Hirai · M. Ichikawa
Chiba University, 1-33 Yayoi-cho, Inage-ku, Chiba-shi, Chiba 263-8522, Japan
e-mail: kmr@faculty.chiba-u.jp

K. Hirai
e-mail: hirai@facutly.chiba-u.jp

© The Author(s), under exclusive license to Springer Nature Singapore Pte Ltd. 2021
H. Kim et al. (eds.), *IT Convergence and Security*, Lecture Notes
in Electrical Engineering 712, https://doi.org/10.1007/978-981-15-9354-3_15

based on the resulting knowledge. Hong et al. developed a system which estimates human's actions (Invasion/Indoor movement) based on the array sensor information [17]. In [17], human's action is estimated by Support Vector Machine (SVM). Tao et al. developed a system which predicts the amount of wind power generation by using deep learning [18].

On the other hand, human emotion estimation is becoming popular for reforming works style, supporting class, and supporting drivers. In [19], students' emotions are analyzed from classroom videos in order to engage students in class. In [19], the emotions analysis is performed by recognizing each student's face expression. From the perspectives from privacy protections, it is desirable to analyze human emotion without camera information. In [20], human emotions analysis method is developed. Since the human emotions are analyzed from vital data in [20], it is necessary for persons to always wear the measurement device.

This paper proposes and develops wireless sensor nodes to monitor working environments and human emotions estimation. The proposed system collects indoor environment data, which concern human perception, via Wireless Sensor Network (WSN). Human emotions are estimated based on the collected environment data. Namely, the proposed system estimates the human emotions without image data form camera sensor or vital data from wearable sensors. Experimental results show the effectiveness of the proposed system.

2 System Structure

Figure 1 shows the structure of the proposed system. The proposed system collects and saves indoor environment data. Sensor nodes measure environment data which concern human perceptions. Sensor nodes send the measured data to the coordinator node. The coordinator node transfers the received data from sensor nodes to the data logger. The data logger logs sensor nodes' data and sends them to the cloud server. Based on the logged data, human emotions are estimated by Machine Learning.

2.1 Sensor-Operation Control

In order to develop measurement equipment, it is necessary to control sensors' operations. Each sensor's operation is controlled by on-board microcomputer. The proposed system uses Arduino as a one-board microcomputer. Each sensor measures the indoor environment data periodically. In this system, sensors measure the temperature, humidity, illuminance, blue light intensity, loudness, odor intensity, human detection, distance, CO_2 concentration, dust concentration, thermography, and atmospheric pressure. Each sensor sends the measured data through a wireless module.

Fig. 1 System structure

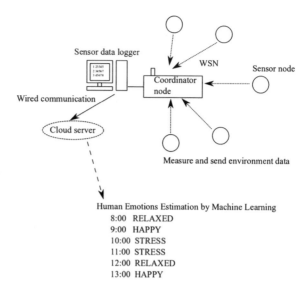

Sensor data logger

WSN

Sensor node

Coordinator node

Wired communication

Cloud server

Measure and send environment data

Human Emotions Estimation by Machine Learning
8:00 RELAXED
9:00 HAPPY
10:00 STRESS
11:00 STRESS
12:00 RELAXED
13:00 HAPPY

2.2 Network Construction

This system uses XBee as a wireless module. A star topology sensor network is constructed. Each sensor sends the measured data to the coordinator node. The coordinator node transfers the received data to the data logger through serial communication. The data logger sends data to the cloud server through wired communication. The logger saves the collected data as csv files. The files include measured data, sensor ID, and the sensor data reception time.

2.3 Sensor Node

This work developed personal sensors which include temperature and humidity sensors, illuminance sensors, blue light intensity sensors, sound sensors, odor intensity sensors, human detection sensors; indoor environment sensors which include CO_2 concentration sensors, dust concentration sensors, and atmospheric pressure sensor; and thermography sensor.

2.3.1 Temperature, Humidity, Illuminance, and Loudness Measurement Sensor

Temperature, humidity, illuminance, and sound sensors get output voltages according to the temperature, humidity, illuminance, and loudness around the sensors. The one-board microcomputer converts the obtained voltages to corresponding temperature (Degree Celsius), humidity (%), illuminance (LUX), and loudness (dB).

2.3.2 Blue Light Intensity Sensor

The developed blue light intensity uses light intensity sensors which **get** output voltage according to the light intensity. In order to measure the blue light intensity, blue color filers were attached on the light intensity sensors. Similar to the illuminance sensor, the one-board microcomputer converts the obtained voltage to the corresponding light intensity (LUX).

2.3.3 Odor Intensity Sensor

Odor intensity sensors get output voltage according the odor intensity around the sensor. The developed odor intensity sensors detect sulfur compound type gas such as hydrogen sulfide and methyl mercaptan. Namely, the developed odor intensity sensors detect offensive odor.

2.3.4 Human Detection Sensor

Human detection sensors are used for detecting persons' motions. Human detection sensor returns 1 if the sensor detects an object. Otherwise the sensor returns 0.

2.3.5 Distance Sensor

Similar to the human detection sensors, distance sensors are used for detecting persons' motions. Distance sensors get output voltage according to the distance between the sensor and objects. The one-board microcomputer converts the obtained voltage to the corresponding distance (cm).

Table 1 Equipped sensors of each node

ID	Equipped sensor
1–14	Temperature and humidity sensor, Illuminance sensor, Blue light intensity sensor, Sound sensor, Odor Intensity sensor, Distance sensor, and Human detection sensor
21–24	Atmospheric pressure sensor, CO_2 concentration Sensor, and dust concentration sensor
31–34	Thermography sensor

2.3.6 CO_2 and Dust Concentrations and Atmospheric Pressure Sensor

CO_2 and dust concentrations and atmospheric pressure measurement sensors get output voltage according to the CO_2 and dust concentrations and atmospheric pressure around the sensors. The one-board microcomputer converts the obtained voltages to the corresponding CO_2 concentration (PPM), dust concentration ($\mu g/m^3$), and atmospheric pressure (hPa). Before developing CO_2 concentration sensors, calibration was performed.

2.3.7 Thermography Sensor

Thermography sensors are used for measuring humans' surface temperature. The proposed system used simplified thermography sensor, which measures temperature around the sensor and sends the measured data (Degree Celsius) as 8×8 pixel image data.

2.4 Measurement System Construction

Environment measurement devices are composed of the developed sensors, one-board microcomputer, and XBee router. Star topology sensor network was constructed in the experimental room. There are one coordinator node and 8 sensor nodes. Table 1 shows the information of the equipped sensors of each node. Each sensor node measures environment every 10 s. Experiment was performed for 60 days.

2.5 Human Emotions Estimation

Human emotions were estimated from obtained indoor environmental data. Human emotions were estimated by Machine Learning method. As the training phase, human emotions were obtained by NEC Emotion Analysis Solution [20]. NEC Emotion

Analysis Solution [20] analyzes four kinds of emotions: HAPPY, ANGRY, SAD, and RELAXED. The proposed system estimates human emotions from the other environmental data. The environmental data (temperature, humidity, illuminance, blue light intensity, loudness, odor intensity, human detection, distance, CO_2 concentration, dust concentration, thermography, and atmospheric pressure) were logged for 60 days. 70% of the logged environmental data were used for training data, and 30% of the logged data were used for test data.

Human emotions were estimated by Random Forest modes, which is one of the supervised Machine Learning based pattern recognition model. We confirmed that the other Machine Learning based model, such as Support Vector Machine and K Nearest Neighbor algorithm, did not achieve enough estimation accuracies. At the training phase, decision tree was created from logged environment data from personal and indoor environment sensors and human emotions data from Emotion Analysis Solutions [19].

3 Experiment

Table 2 shows the emotions estimation accuracy of a person. The human emotions were estimated from 12 kinds of sensors. It is seen from Table 2 that the emotions estimation accuracy was over 80%.

Table 3 shows the importance of the obtained data. It is seen from Table 3 that the importance of the CO_2 concentration was the highest among 12 kinds of sensors. It is also seen from Table 3 that the importances of distance and thermography were relatively high for estimating human emotions.

Table 4 shows the estimation accuracy of human emotions. Human emotions were estimated by Random Forest model. Random Forest model is one of the supervised Machine Learning based pattern recognition model. It is seen from Table 4 that the estimation accuracy differs by the kinds and the number of input sensor data. Using multiple kinds of sensors improves the estimation accuracy. It is seen from Table 4 that the estimation accuracy was over 80% by using more than 9 kinds of sensors. It is also seen from these results that WSN based big data collection is useful for human

Table 2 Ratio of each emotion and estimation accuracy of a person

Number of data	14420
Happy raito	0.666
Stress ratio	0.221
Relaxed ratio	0.095
Sad ratio	0.018
Train accuracy	0.999
Estimation accuracy	0.805

Table 3 Importance of each sensor

Sensor data	Importance
Temperature	0.019
Humidity	0.014
CO_2 concentration	0.149
Illuminance	0.080
Blue light intensity	0.085
Loudness	0.063
Odor intensity	0.073
Distance	0.101
Human detection	0.039
Dust concentration	0.062
Atmospheric pressure	0.068
Thermography	0.103

Table 4 Emotions estimation accuracy versus the number of kinds of sensors

Number of kinds of sensors	Person 1	Person 2	Person 3
1	0.606	0.530	0.710
2	0.638	0.630	0.755
3	0.758	0.752	0.821
4	0.785	0.790	0.835
5	0.805	0.789	0.821
6	0.806	0.788	0.822
7	0.808	0.794	0.820
8	0.813	0.786	0.810
9	0.822	0.803	0.832
10	0.810	0.787	0.827
11	0.808	0.787	0.823
12	0.799	0.780	0.827
13	0.805	0.799	0.828
14	0.805	N/A	N/A

emotions estimation. It is seen from the experimental results that the developed personal and indoor environment sensors are helpful to estimate human emotions.

4 Conclusion

This paper proposed and developed wireless sensor nodes to monitor working environments and human emotions estimation. The proposed system collects indoor environment data, which concern human perception, via WSN. Human emotions were estimated based on the collected environment data. Namely, the proposed system estimated the human emotions without image data form camera sensor or vital data from wearable sensors. Experimental results showed that the proposed system achieves over 80% estimation accuracy by using multiple kinds of sensors, which showed the effectiveness of the proposed system. Future works include the investigation of the number of sensor kinds and the improvement of the human emotions estimation accuracy.

Acknowledgements This work has been supported by Innovation Platform for Society 5.0 at MEXT.

References

1. Tseng SM (2003) A high-throughput multicarrier DS CDMA/ALOHA network. IEICE Trans Commun E86-B(4):1265–1273
2. Komuro N, Habuchi H, Kamada M (2004) CSK/SSMA ALOHA system with nonorthogonal sequences. IEICE Trans Fund E87-A(10):2564–2570
3. Komuro N, Habuchi H (2005) A reasonable throughput analysis of the CSK/SSMA unslotted ALOHA system with nonorthogonal sequences. E88-A(6):1462–1468
4. Sekiya H, Tsuchiya Y, Komuro N, Sakata S (2011) Analytical expression of maximum throughput for long-frame communications in one-way string wireless multihop networks. Wireless Pers Commun 60(1):29041
5. Komuro N, Habuchi H, Tsuboi T (2008) Nonorthogonal CSK/CDMA with received-power adptive access control scheme. IEICE Trans Fund, E91-A(10):2779–2786
6. Chavan SD, Kulkarni AVK (2019) Improved bio inspired energy efficient clustering algorithm to enhance QoS of WSNs. Wireless Pers Commun 109(3):1897–1910
7. Lindsey S, Raghavendra C, Sivalingam KM (2002) Data gathering algorithms in sensor networks using energy metrics. IEEE Trans Parallel Distrib Syst 13(9):924–935
8. Fan X, Song Y (2007) Improvement on LEACH protocol of wireless sensor network. In: Proceedings International Conference on Sensor Technologies and Applications (SENSOR-COMM), 260–264
9. Luo CY, Komuro N, Takahashi K, Kasai H, Ueda H, Tsuboi T (2008) Enhancing QoS provision by priority scheduling with interference drop scheme in multi-hop Ad Hoc network. IEEE Global Communication Conference (GLOBECOM), 1321–1325 (2008).
10. Tuan DT, Sakata S, Komuro N (2012) Priority and admission control for assuring quality of I2V emergency services in VANETs integrated with wireless LAN mesh networks. International Conference on Communications and Electronics (ICCE), 91–96
11. Sony CT, Sangeetha CP, Suriyakala CD (2015) Multi-hop LEACH protocol with modified cluster head selection and TDMA schedule for wireless sensor networks. Proceedings Global Conference on Communication Technologies (GCCT), 539–543
12. Plageras AP, Psannis KE, Stergiou C, Wang H, Gupta BB (2018) Efficient IoT-based sensor big data collection-processing and analysis in smart buildings. Future Gener Comput Syst 82:349–357

13. Kelly SDT, Suryadevara NK, Mukhopadhyay SC (2013) Towards the Implementation of IoT for environmental condition monitoring in homes. IEEE Sens J 13(10):3846–3853

14. Byun J, Jeon B, Noh J, Kim Y, Park S (2012) An intelligent self-adjusting sensor for smart home services based on ZigBee communications. IEEE Trans Consum. Electron. 58(3):794–802

15. Gill K, Yang S-H, Yao F, Lu X (2009) A ZigBee-based home automation system. IEEE Trans Consum Electron 55(2):422–430

16. Weixing Z, Chenyun D, Peng H (2012) Environmental control system based on IoT for nursery pig house. Trans Chin Soc Agric Eng 28(11)

17. Hong J, Ohtsuki T (2011) A state classification method based on space-time signal processing using SVM for wireless monitoring system. Proc IEEE PIMRC

18. Tao Y, Chen H, Qiu C (2012) Wind power prediction and pattern feature based on deep learning method. National Program on Key Basic Research Project (973 Program) under Grant 2012CB215201

19. Zeng H, Shu X, Wang Y, Wang Y, Zhang L, Pong TC, Qu H (2020) EmotionCues: Emoion-oriented visual summarization of classroom videos. IEEE Trans Visual Comput Graph Early Access

20. NEC Emotion Analysis Solution (2018). https://jpn.nec.com/press/201806/20180611_01.html

Part V
Security and Privacy

Analysis of Automatic Access Monitoring and Control Systems and Anti-Hacking Systems

Egor Efremov, Alex Kovalevsky, and Maria Skvortsova ⓘ

Abstract This paper discusses the vulnerabilities of automatic access monitoring and control systems widely used in everyday life. Two basic principles of information exchange in such devices are described. Study of amplitude-frequency characteristics of key fobs from various manufacturers of automatic access monitoring and control systems was undertaken. According to the results of the study and the vulnerabilities found, a universal code grabber model is shown. Knowing the principles of operation of this model, it will be possible to successfully counter hacking. A test model of the system based on the Arduino Uno microcontroller was also developed.

Keywords Anti-hacking · Car alarm systems · Automatic systems hacking · Monitoring system · Control system · Arduino uno · Vulnerability

1 Introduction

At present, when scientific and technological progress takes seven-league steps forward, computer systems are taking up more and more space in our lives. Many people store important documents on their electronic devices (smartphones, computers, tablets), transfer confidential data via open communication channels, trust the security of their property not to mechanical but to computer information protection systems. However, few people think how secure is the information and those applications where sensitive data is stored. According to world studies, over the past 5 years, personal data has been hacked and stolen alternately from multimillion companies (these include financial institutions, search engines, cloud storages). Some states abuse their rights and also, through legal and illegal means, gain access to personal information of residents. Therefore, this study is relevant and is dedicated to the search for vulnerabilities of the automatic access control system manufactured by Came, as well as the development of recommendations for countering unauthorized access to objects protected by these systems.

E. Efremov (✉) · A. Kovalevsky · M. Skvortsova
Bauman Moscow State Technical University, Moscow 105005, Russian Federation
e-mail: e.efremov2014@gmail.com

© The Author(s), under exclusive license to Springer Nature Singapore Pte Ltd. 2021   161
H. Kim et al. (eds.), *IT Convergence and Security*, Lecture Notes
in Electrical Engineering 712, https://doi.org/10.1007/978-981-15-9354-3_16

2 Theoretical Description

The access monitoring and control system (AMCS) is a combination of software and hardware, organizational and methodological tools that solve the problems of controlling access to the room. AMCS protects against intruders, allows to organize internal security and improve discipline at the facility [1, 7, 8, 9].

Currently, the access monitoring and control system is in demand at almost any facility. Even the most minimal access restriction, based on electromagnetic or electromechanical locks, already plays a significant role in maintaining order and safety on the object. Limiting entry to outsiders is one of the main tasks of AMCS.

To understand the principle of device operation and to study its properties, the AMCSs from the following manufacturers were analyzed: Came [2], Nice [3], Doorhan [4], Bft [5] and Comunello [6]. Despite the variety of AMCSs manufacturers and techniques to install and use them, there are two basic principles of information exchange in such systems: static and dynamic. When the receiver and transmitter of the signal exchange static information, the use of a signal replay attack allows to bypass the access control system. The considered scheme for obtaining unauthorized access based on the Came security system is implemented in similar ways at urban infrastructure facilities, corporate facilities and private property. The use of non-standard frequencies for signal transmission, dialogue between the receiver and the transmitter and other techniques can increase the security level of access control systems, but does not guarantee absolute security of the protected object [8, 9, 10].

3 Information About the Structure and the Principle of Operation of Access Control Systems from Different Manufacturers

The considered automatic access monitoring and control system Came [2] is a system of two-channel radio receiver of control commands operating at a frequency of 433.92 MHz and key fob transmitters at a frequency of 433.92 MHz with an operating range of about 20 m. Below is a snapshot from a frequency scanner for this device (Fig. 1).

As can be seen in the figure, the frequency range 432.25–434 MHz is monitored, the corresponding pictures are observed in the spectrogram (Fig. 2) and sonogram (Fig. 3) while transmitting the signal.

To add a new transmitter to the receiver's memory, it is necessary to start the operation of copying the code from one key fob to another. Therefore, there is no need to make any manipulations with the receiver. We consider the transmitted signal in more detail.

The static code format is still relevant. Formats of different companies differ from each other by pulse durations. To create the transmitters, Holtek HT-12E coder integrated circuits are used—in this case, microswitches are installed in the key fobs,

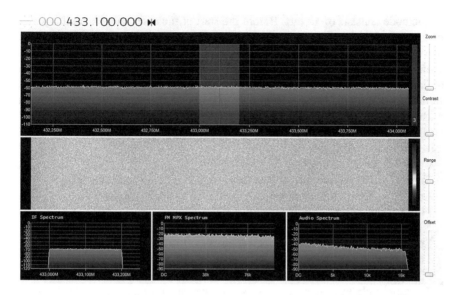

Fig. 1 A figure caption is always placed below the illustration. Short captions are centered, while long ones are justified. The macro button chooses the correct format automatically

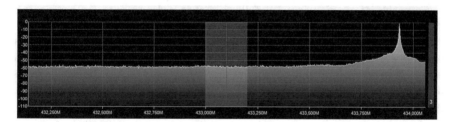

Fig. 2 The moment of signal transmission on the spectrogram

Fig. 3 The moment of signal transmission on the sonogram

which sets the key fob number. They may have one or more buttons—the coder IC allows to implement from 1 to 4 buttons. Pulse durations are set by a resistor in the required components of the coder. There are coders without a key pad, the serial number is flashed into the IC by the manufacturer.

The code consists of 12 bits. Before the start of the code, there is a pilot period (consisting of 36 low-level intervals) and a start pulse (consisting of 1 high-level interval). It is the duration of the start pulse that indicates which time interval is used when generating the signal. After the start bit, the code itself begins, the start of the bit is from a low level. After the end of the last pulse, the code word is repeated starting from the pilot period. With a short press on the button, 4 code words are transmitted, while holding the button, the codes are cyclically repeated (in series of 4 code words). The possible number of code combinations is 4096. We consider the pulses of various AMCSs.

Came [2] pulse durations:

Log "1"—640 µs low level (two intervals), 320 µs high (one interval).

Log "0"—320 µs low level (one interval), 640 µs high (two intervals).

Pilot period—11,520 µs, start pulse—320 µs.

NICE [3] pulse durations:

Log "1"—1400 mks low level (two intervals), 700 mks high (one interval).

Log "0"—700 mks low level (one interval), 1400 mks high (two intervals).

Pilot period—25,200 µs, start pulse—700 µs.

The amplitude-frequency characteristic of the signals of the Came/Nice packet is shown in Fig. 4.

Other pulse durations are possible, but the ratio of high and low levels, as well as the pilot period and the start bit, are formed according to the same principle.

From the information found it follows that in order to open the barrier, it is necessary either to copy the code at the time of its transmission, or to guess it.

In the course of the study, the amplitude-frequency characteristics of the key fobs of various companies producing various AMCSs were investigated, including car alarms. As a result, it became known that the vast majority of AMCS manufacturers use the same frequency, namely 433.92 MHz. However, there are companies, in particular Starline, which uses a frequency of 434.5 MHz, and the signal is sent pulsed and very quickly. This causes great difficulty in intercepting such signals due to the non-standard frequency and short transmission time.

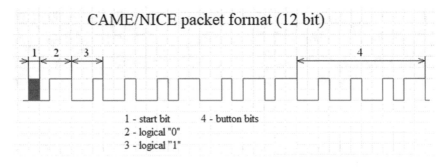

CAME/NICE packet format (12 bit)

1 - start bit   4 - button bits
2 - logical "0"
3 - logical "1"

Fig. 4 The amplitude-frequency characteristic of the signal

4 Building the AMCS Test Model Based on the Arduino Uno Microcontroller

Arduino Uno [13] was chosen as a platform for developing a test model—it is an open platform that allows to assemble all kinds of electronic devices [14]. The platform consists of hardware and software. For programming, a simplified version of C++ is used, also known as Wiring.

In addition to Arduino, the following were used:

- 433 MHz Wireless Transmitter (WRL-10534) [14].
- 433 MHz Wireless Receiver (WRL-10532) [15].

The result of this phase of the research work was the creation of a prototype that includes models of the devices "barrier–key fob" and "barrier–interceptor–key fob" (Figs. 5 and 6).

The model "barrier–key fob" is the interaction of two modules: a receiver that simulates a security device and a transmitter that simulates the operation of a key fob (a key fob from a conventional security system operating at a frequency of 433 MHz can be used).

The model "barrier–interceptor–key fob" consists of the above modules and the "interceptor". The "interceptor" is a device that in the background scans and stores in the internal memory all signals at a frequency of 433 MHz.

The principle of operation of the model "barrier–interceptor–key fob" is described below.

First of all, it should be noted that this model "barrier–interceptor–key fob" (universal code grabber model) is suitable for various automated security systems,

Fig. 5 Transmitter

Fig. 6 Receiver

as well as for some models of car alarms (Fig. 7). It is understood that the key is not static, but constantly changing, i.e. dynamic. With a static key, everything is much simpler—just intercept it once. Hacking such a security system can be done in different ways, for example, by writing a large number of keys, you can try to figure out the key generation algorithm, but this method does not promise success. The most interesting is the following hacking model. The hacker waits for a person with a key fob from the security system and then for the start of the signal transmission. Then, after sending the signal, the code grabber records the signal, but at the same time the air is blocked out in the direction of the protected object, thus the signal does not reach the object. Naturally, a person will try to press the button again to repeat the signal transmission. At this moment, we carry out similar events, but with the addition of the following: as soon as we get the second key, jamming stops, for example after 500 ms, and from our code grabber we send a signal with the first key.

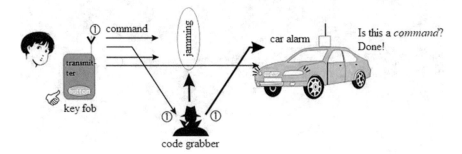

Fig. 7 Dialogue system diagram

As a result, we have one valid key. The advantages of this method are that it does not matter to us how long the key is and by what algorithm it is generated [11, 12, 16].

5 Conclusion

According to the study, the following recommendation was developed, according to which, for reasons of counteracting such code grabber models, it is proposed to use time protection in conjunction with a "dialogue". It is only worth adding to this that if no answer is received within a certain number of milliseconds, then the dialogue is considered failed and will start again.

Currently, as part of the study, the development of a device capable of analyzing the protection levels of specific AMCS by directly compromising the system continues. Based on the results of testing the AMCSs, it will be possible to identify vulnerabilities of a mass nature, as well as to find the most successful implementations to counter intrusions into the operation of security systems. The result of these measures is the development of the most secure AMCS prototype to demonstrate resistance to various types of attacks on the system.

References

1. ACS analysis, https://blog.terra-security.ru/analiz-skud-kak-proverit-sistemu-kontrolya-i-upr avleniya-dostupom-na-effektivnost. Last accessed 4 Feb 2020
2. Official web-site of access control system manufacturer Came, https://camerussia.com/. Last accessed 4 Feb 2020
3. Official web-site of access control system manufacturer Nice, https://niceforyou.ru/. Last accessed 4 Feb 2020
4. Official web-site of access control system manufacturer Doorhan, https://doorhan.ru/. Last accessed 4 Feb 2020
5. Official web-site of access control system manufacturer Bft, https://www.bft-automation.com/. Last accessed 4 Feb 2020
6. Official web-site of access control system manufacturer Comunello, https://comunello.ru/. Last accessed 4 Feb 2020
7. Kozlova NY, Fomichev AV (2019) Development of the remote control system of the free-flying space manipulation robot with force feedback. In: AIP conference proceedings, vol 2171, no 1. AIP Publishing LLC, p 150005
8. Buldakova TI, Sokolova AV (2019) Network services for interaction of the telemedicine system users. In: 2019 1st international conference on control systems, mathematical modelling, automation and energy efficiency, SUMMA. IEEE, pp 387–391
9. Youness SF, Lobusov EC (2019) Networked control for active suspension system. Procedia Comput Sci 150:123–130
10. Repp P (2018) Theoretical aspects of cyber-attack modeling. In: 2018 International Russian automation conference, RusAutoCon. IEEE, pp 1–5
11. Balepin I, Maltsev S, Rowe J, Levitt K (2003) Using specification-based intrusion detection for automated response. In: International workshop on recent advances in intrusion detection. Springer, Berlin, Heidelberg, pp 136–154

12. Gurenko VV, Bychkov BI (2019) The discretization of the energy characteristics of signals in harmonic simulation algorithms. In: 2019 IEEE conference of Russian young researchers in electrical and electronic engineering, EIConRus. IEEE, pp 2142–2147
13. Official web-site of the Arduino platform, www.arduino.cc, last accessed 2020/02/04
14. Transmitter TWS-BS-3 technical documentation, https://cdn.sparkfun.com/datasheets/Wir eless/General/TWS-BS3_433.92MHz_ASK_RF_Transmitter_Module_Data_Sheet.pdf, last accessed 2020/02/04
15. Receiver TWS-BS-3 technical documentation, https://www.sparkfun.com/datasheets/RF/ KLP_Walkthrough.pdf, last accessed 2020/02/04
16. Krasovsky A, Vasyukov S, Murzin I (2019) Electrical model of car power wiring for transmission of pulse control signals through it. In: 2019 International Russian automation conference, RusAutoCon. IEEE, pp 1–5

Part VI
Other Related Topics

An Adaptive Low Power Schedule for Wireless Sensor Network

Almamoon Alauthman, Wan Nor Shuhadah Wan Nik, and Nor Aida Mahiddin

Abstract A Wireless Sensor Network (WSN) is used in many applications such as remote monitoring and tracking, healthcare, industrial settings, automated and self-adjusting systems at homes and factories. In these systems, energy consumption is a major concern as most devices in the network rely on small batteries for energy source. In such cases, minimization of energy usage can prolong the lifetime of sensor's battery. Therefore, this paper proposed a strategy to improve the efficiency of power consumption in WSN by decreasing the number of idle listening state which further reduces power consumption. By doing this, it is expected that network life time can be prolonged. Undoubtedly, large amount of energy is wasted in idle listening state. We intelligently improved the original TPO technique in order to achieve less power consumption. The experiments is conducted by using Omnet++ 4.6 with Mixim library and the results showed that when compared with the original TPO technique, the proposed technique achieves significant energy saving which is up to 39.3% better than the original TPO technique.

1 Introduction

Energy efficiency and the low power consumption is one of the major considerations for WSN. As sensors in WSN are normally powered by batteries, it is crucial to conserve power as much as possible to maximize the WSN lifetime [1]. A Time-Division Multiple Access (TDMA) is an attractive MAC protocol for efficient data collection in WSN [2]. It is essential to reduce power consumption at the sensor node

A. Alauthman (✉) · W. N. S. W. Nik · N. A. Mahiddin
Faculty of Informatics and Computing, Universiti Sultan Zainal Abidin, Besut Campus, Terengganu, Malaysia
e-mail: Almamoon.athamneh1@gmail.com

W. N. S. W. Nik
e-mail: wnshuhadah@unisza.edu.my

N. A. Mahiddin
e-mail: aidamahiddin@unisza.edu.my; Almamoon.athamneh1@gmail.com

© The Author(s), under exclusive license to Springer Nature Singapore Pte Ltd. 2021
H. Kim et al. (eds.), *IT Convergence and Security*, Lecture Notes in Electrical Engineering 712, https://doi.org/10.1007/978-981-15-9354-3_17

to maximize the network lifetime of WSN for data collection [3]. Several reasons caused wasted energy, overhearing, collisions, and idle listening, which is the most relevant to our work. This paper proposes an improved TDMA schedule that achieves high energy efficiency for tasks in WSN to maximize network lifetime. Our proposed model enables each node in WSN to receive the data packet from its children without any additional idle listening state. The experimental results showed that the proposed technique significantly improves the energy efficiency of sensor nodes as compared to the original TPO technique. The organization of this paper is as follows. Section 2 presents the related work. Section 3 shows the system model. Section 4 describes the proposed technique. Section 5 analyzes the performance of the proposed technique. The simulation result is presented in Sects. 6 and 7 concludes the paper.

2   Related Work

TDMA scheduling for sensor data collection in WSN has attracted much research attention in recent years. Some early work aims to construct schedules for each node to communicate once with each of its neighbours [4]. Authors in [5] presented a slot allocation algorithm based on TDMA. In this research, slots in TDMA are arranged based on local information at each sensor nodes. Where slots are divided for packet transmission and acknowledgment purpose. Collisions are avoided as each node is provided with different slots. Further in [6], another TDMA based protocol is proposed to decrease energy consumption in WSN. However, the proposed model suffers a negative effect on the overall throughput of the system. Another MAC protocol named Energy FDM based on Carrier Sense Multiple Access (CSMA) is proposed in [7]. The algorithm reduces energy consumption by transmission power control, but it leads to a high collision rate, especially in high traffic networks. Further, the pTunes which proposed in [8] saves energy by adapting its own MAC parameters on the channel models, but the network information collection scheme results in transmission delay. Considering the defects of the protocols above, on the other side, some hybrid scheme protocols are proposed to save energy in WSN. In [9], the authors proposed a novel TDMA schedule that is efficiently collecting sensor data for network traffic pattern called Traffic Pattern Oblivious (TPO). In this schedule, the energy consumed by sensor nodes for any traffic pattern is proved to be low. In [10], the author introduces extra-bit technique to reduce idle listening. In this research, each packet includes an extra bit which informs the receiver node on whether or not further data packets will follow. In [11], the authors proposed an efficient routing protocol in order to extend the network lifetime for continuous data collection with dynamic traffic patterns by the construction of a Directed Acyclic Graph (DAG).

3 System Model

Consider a sensor network in a tree topology. The tree network is represented as a graph $G = (N, E)$, where N represents sensors and E represents communication links between sensors, and the sink or the base station is the root of the tree. Assume that all nodes have a unidirectional transceiver for monitoring some application or area. All communication among sensors are performed over a single frequency channel, where nodes cannot send and receive a packet at the same time, By considering a TDMA schedule, time is divided into a number of slots of equal length and several packets in each slot can be scheduled. Data are collected from sensors in order to minimize the number of TDMA time slots which further minimizes the power consumption and prolonged the WSN lifetime. Assume that not every sensor has a data to be collected in each round of data collection. If a transmission from node N to its parent P is scheduled in a particular time slot but there is no data at N to be sent to P, then the idle listening will occurred. At this time, P will know that there is no data that will be sent from N in this round. Therefore, the transceivers of P can be switched off and further saves the energy as no power is consumed.

4 A Modified TPO (MTPO) Technique

In the TPO schedule which are proposed in [9], all transmissions from a node to its parent must occur in successive slots, starting from the first slot of TDMA, with a non-empty slot in a TDMA between two packets, to reduce the idle listening state. The main advantage of TPO scheduling is that as soon as the parent node finds one empty slot, it knows that there will be no more data from its child will arrive at this round onwards. Consequently, the parent node will switch its transceiver off in order to avoid idle listening state, and save power. Generally, a node must listen to transmissions from its child nodes until an idle state happen, to indicate the end of transmission from that child, and further allows the parent to switch off its transceiver. Each time the parent has listened to all of its child and has received at least one data packet, it can make one transmission to its own parent. This process continues until it has received idle listening slot from its entire child. Finally, it will transmit the remaining packets to its own parent. This process continues for all nodes until the base station node has stopped listening to its entire child, implying that all packets have been received. The procedure of the proposed technique which is a modification of TPO technique (MTPO) is as follow:

- In the last level of tree (LN) a TPO schedule is applying in every nodes, all parent for this level make TDMA schedule with slots equal to the number of its children.
- At level N-1, when the parent find an empty slot in the TDMA schedule, meaning that an idle listening state happen, For that the parent knows there is no data packet will received at this round onwards, then switched off its receiver and computing the number of data packet received.

- Every parent at Level N-1, send the number of packet received in the previous step as message to the upper parent which located at level N-2, so the parent at level N-2 knows in advance the number of data packet on its child.
- Every parent at level N-2 make a TDMA schedule with a number of slot equal to the number of data packet number which received in the previous step, and then switch off its transceiver after the number of packed is receive, for that there is no idle listening state happen at every levels between Level N-1 and the sink level.

The difference between the proposed schedule and TPO schedule, is that the TPO schedule take one slot more than the number of packet to know that this is the final data packet and there is no more any packet may be coming and then switch off it receiver, while the proposed schedule said that the parent of last row in the tree is only parents that take one slot more than the number of packet and the others level have no any idle listening state.

5   Expected Result

The communication energy cost, Ecomm, is the most important factor of the operational costs in WSN. The Ecomm component are transmission, receiving, and listening, as show in the following equation:

$$Ecomm = Etx + Erx + Elisten \tag{1}$$

The transmission energy Etx, refers to the energy consumed during the packets transmission, Erx refers to the energy consumed when receiving packets, and Elisten is the radio energy consumption when the radio is active, but not receiving or sending packets. This listening is to check for messages on the radio channel

$$Elisten = Vdc * Ilisten * Tlisten \tag{2}$$

where Ilisten is the current draw of the radio in listening mode, Tlisten is the number of idle listening state, which depends on the MAC protocol [12], for that if we eliminated every idle listening slot on TDMA, the Elisten is reduced, power consumption is also reduced, then the total life time of WSN is maximized, which is the objective of our study.

6   Simulation Result

In the simulated network, sensor nodes were randomly deployed in a 1×1 rectangular area, with the base station situated in the centre. The radio propagation ranges of different sensor nodes were initialized to 0.25. We conducted experiments over many random deployed networks of different sizes, and obtained similar experimental results. For simplicity, we shall primarily present the performance trends for a sample network of 32,64,256 and 1024 sensors, which is varying between small, medium, and large network. The base station was equipped with unlimited energy supply, energy budget allocated to each sensor node was identical. The energy units spent by a sensor node on sending and receiving (listening) in a time slot were initialized to be 1 and 0.75. We create a Network and calculating the energy for all nodes, select the cluster head, based on the energy level and make a cluster. The cluster head act as a parent node, the cluster member act a child node, and the base station act as a root. Next make a scheduling based on TPO and MTPO, the idle nodes are go to sleep nodes based on the scheduling. We generate the graph for Energy Consumption versus number of Sensors. We use the simulation tool Omnet++ 4.6 with MIXIM library.

6.1   Power Consumption

Wireless sensor network life time depend in the battery of each sensor. This research focuses on minimizing power consumption to maximize the network lifetime. Figure 1 shows the average power consumption for TPO and MTPO under different number of sensors, i.e. 32, 64, 256 and 1024, Each graph shows the power consumption, as shown the power consumption varies depending on the number of sensors, MTPO consume less power than TPO by 33.7%.

Table 1 shows the difference power consumption between TPO and MTPO, as shown in the table the power consumption of MTPO is significantly better than TPO in large network by 33.7%, and for small network by 16.8%.

7   Conclusion

A modified TDMA schedule based on TPO schedule is presented in this paper, the proposed technique achieves high energy efficiency, In this technique, the energy consumed by sensor nodes is very close to the minimum, by reducing the number of idle listening state in a data collection rounds. We have conducted simulation experiments of approximate data collection using Omnet++ simulation program. The simulation results show that compared with traffic pattern oblivious algorithm, the proposed technique achieves significant energy savings.

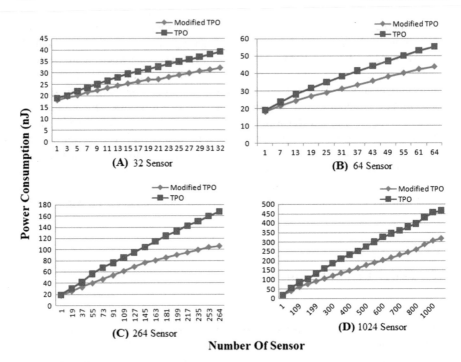

Fig. 1 Power consumption comparison using TPO versus MTPO

Table 1 Power consumption comparison between TPO and MTPO

Number of sensor	MTPO	TPO	Parentage of improvement
12	59	72	16.8
16	60	77	22
20	64	82	24.3
26	69	88	25
32	73	97	25.6
64	79	108	26.8
128	103	142	28.1
256	139	201	29.8
512	209	300	30.3
1024	331	492	33.7

References

1. Werner-Allen G, Lorincz K, Ruiz M, Marcillo O, Johnson J, Lees J, Welsh M (2006) Deploying a wireless sensor network on an Active Volcano. IEEE Internet Comput 10(2)
2. Ma J, Lou W, Wu Y, Li X, Chen G (2009) Energy efficient TDMA sleep scheduling in wireless sensor networks. In: Proceedings of the IEEE INFOCOM '09, pp 630–638
3. Rault T, Bouabdallah A, Challal Y (2014) Energy efficiency in wireless sensor networks: a top-down survey. Comput Netw 67(Supplement C):104–122
4. Gandham S, Dawande M, Prakash R (2005) Link scheduling in sensor networks: distributed edge coloring revisited. In: Proceedings of the IEEE INFOCOM '05, pp 2492–2501
5. De Guglielmo D, Anastasi G, Conti M (2013) A localized slot allocation algorithm for wireless sensor network. Ad Hoc networking workshop (MED-HOC-NET). IEEE, Ajaccio
6. Amokrane A, Langar R, Boutaba R, Pujolle G (2012) Agreen framework for energy efficient management in TDMAbased wireless mesh networks. In: Proceedings of the 8th IEEE international conference network and service management, vol 62. IEEE, Las Vegas, NV, USA, pp 322–328
7. Al-Kadri MO, Aijaz A, Nallanathan A (2016) An energy- efficient full-duplex MAC protocol for distributed wireless networks. IEEE Wirel Commun Lett 5(1):44–47
8. Zimmerling M, Ferrari F, Mottola L, Voigt T, Thiele L (2012) pTunes: runtime parameter adaptation for low-power MAC protocols. In: Proceedings of the 11th ACM/IEEE conference on information processing in sensing networks (IPSN '12). Beijing, China, pp 173–184
9. Zhao W, Tang X (2013) Scheduling sensor data collection with dynamic traffic patterns. IEEE Trans Parallel Distrib Syst 24(4):789–802
10. Rasul A; Erlebach T (2017) The extra-bit technique for reducing idle listening in data collection. Int J Sensor Netw (IJSNET) 25(1)
11. Zhao W, Tang X, Xu L (2018) Constructing routing structures for sensor data collection with dynamic traffic patterns. Comput Netw
12. Razzaque MA, Dobson S (2014) Energy-efficient sensing in wireless sensor networks using compressed sensing. In: Proceedings of the 14th ACM/IEEE international symposium on low power electronics and design. Sheraton La Jolla, CA, USA, 11–13 August 2014

A Reduction Model for Simulating Large-Scale Interconnection Network

Tien Thanh Nguyen⬤, Cuong Van Bui⬤, and Khanh-Van Nguyen⬤

Abstract Most popular network simulation tools share some common limitations, including their execution time being too long that undermines performance of network topologies, routing algorithms and related issues. This paper proposes a reduction model for simulating packet transfer process in large-scale interconnection networks. Our proposed simulation framework relies on the reduction of packet transfer process, deterministic finite automata for formal specification and flexible architecture for implementation. Our reduction model supports scalability for large-scale interconnection network and its evaluation highlights the advantages of network design.

Keywords Network simulation · Reduction model · Large-scale interconnection network

1 Introduction

As computers and telecommunication technology become cheaper and the growing demand for data storage, data centers can evolve to very large sizes such as tens of thousands of nodes or more. These data centers need innovative designs of effective interconnection topologies, routing algorithms and a congestion control mechanism to achieve the best performance. In order to design a new interconnection topology with such a large size, researchers would need to make use of network simulations as soon as possible so as to evaluate the performance factors during the early stages of the designing process.

T. T. Nguyen (✉) · C. Van Bui · K.-V. Nguyen
Hanoi University of Science and Technology, Hanoi, Vietnam
e-mail: nguyenthanh@soict.hust.edu.vn

C. Van Bui
e-mail: cuongbv765@gmail.com

K.-V. Nguyen
e-mail: vannk@soict.hust.edu.vn

© The Author(s), under exclusive license to Springer Nature Singapore Pte Ltd. 2021   179
H. Kim et al. (eds.), *IT Convergence and Security*, Lecture Notes
in Electrical Engineering 712, https://doi.org/10.1007/978-981-15-9354-3_18

However, the most popular network simulations such as NS3 [1], Omnet++ [2], and JiST [3] can be seen as too slow to evaluate the performance of these large-size interconnection topologies, and in particular to experiment with their custom routing algorithms and congestion control strategies. This is because these popular simulation tools are built to evaluate network protocols under varying network conditions [4]. The complete implementation of a full stack network protocols dramatically slows down execution time of the network simulation, reduces throughput and increases latency. When used for simulating interconnection networks, these popular simulations thus undermine the performance of network topologies, routing algorithms, and congestion control. On the other hand, when we analyze an interconnection network solution, we focus on the characteristics of the topology, the logic of routing policy and the distribution of traffic, wherein the technical characteristics of how data transmission is performed per a fixed, given route is not relevant. In other words, we need the simulation to do the real stuff only to analyze the shape and routing rule of the network; hence, the lower-level techniques such as error detection and correction could be relaxed or even nullified (as in an ideal transmission environment).

This situation leads to the desire of a high-level abstraction of network simulation which helps focus on the higher levels and thus, helps design a new network simulation tool specialized for experimenting with interconnection networks. There are several researchers who develop their own network simulations to evaluate the performance of their proposal network topologies, routing algorithms and congestion control strategies. These specialized tools have higher-level abstraction than the standard simulation tools (NS3, Omnet++, JiST), but they all lack a formal specification for their reduction models. This shortage makes it difficult to validate these models and raises a question about the accuracy of evaluation. In addition, the shortage leads to difficulties for software maintenance and evolution in the future.

Our paper proposes a reduction model for simulating communication process (such as packet transfer) in large-scale interconnection networks. Our reduction model has a high-level abstraction. This abstraction allows our model to focus on the generation of packets, as well as, sending, forwarding and receiving them at source nodes, intermediate nodes, and destination nodes. Focusing on analyzing topology and routing characteristics, our model omits high layers above the Network layer and simplifies behaviors of the three lowest layers. This paper also introduces a formal specification for our reduction model and a flexible framework to implement this model.

In this paper, Sect. 2 introduces related work; Sect. 3 describes our proposed reduction model and formal specification of this model. Section 4 elaborates on the implementation and section V is the conclusion and future work.

2   Related Work

Most popular network simulations are discrete event systems (DES) [5]. All entities in network simulations, such as source nodes, intermediate nodes, destination nodes as well as links, are generators of ongoing discrete events. The network simulations arrange these events according to the time when events are executed. The execution of events, in turn, changes states of entities and leads to many more events in the future. This discrete event model is a de facto standard for building network simulations. Among these popular simulations, NS3 is the most efficient simulation tool [6].

However, NS3 mirrors real network components, protocols, and APIs closely. This detailed implementation slows down the simulating process (for example, due to time spent on serializing and de-serializing packets [7]) as well as undermines the performance of network topologies, routing algorithms and congestion control. Three main reasons for this undermining are: (i) NS3 suffers artificial congestion caused by its attempt to establish internal sockets to send/receive packets between layers; (ii) nodes spend too much time on serialization and de-serialization of numerous large packets; (iii) NS-3 involves limited size queues on each network layer that can cause congestion. These reasons increase latency and reduce throughput in many test cases.

Some authors implement their proposed network topologies and/or routing algorithms by using physical machines instead of any simulation. Given prototypes, researchers could get an accurate insight into the system behavior, performance, and implementation bugs. For example, [8] by setting up several prototypes of some network topologies in their computers, they can evaluate throughput of their proposals. This methodology has empirical results but, in general, it suffers limitation in scalability for larger-scale interconnection networks.

Some authors also notice that existing network simulations are not suitable for larger-scale networks because these simulations could take hours to simulate just one test case [9]. Therefore, they build their own simulation tools that allow users to implement both network topologies as well as routing algorithms. However, these tools have no formal specification which is crucial for modeling, observation and control DES [10].

Based on mathematical models, some analytical systems have the highest level of abstraction. These tools rely on numerical solver to yield fast evaluation results, but they usually focus only on the relevant aspects of a network and ignore the irrelevant ones. For example, the tool of [11] restricts researchers to implement their own routing algorithms and congestion control.

3 Reduction Model and Specification

3.1 Our Proposed Model

Considering several advantages and disadvantages of the above mentioned methodologies, we propose a reduction model for simulating interconnection networks with the focus on analyzing routing characteristics, network throughput and related issues. As with other simulation models, our proposed model is a discrete event system [5] but unlike the popular simulators, ours does not offer the full OSI 7-layer protocol

Fig. 1 Source queue and exit buffer of a source node (left). Entrance buffers and exit buffers of switch has k = 2 links (right)

stack. Our model focuses on providing the three lowest layers: Network layer, Data Link layer, and Physical layer. In our proposed model, the Network layer has the only one role: for doing the routing of packets. Meanwhile, the Data Link layer is responsible for only congestion control strategies as well as updating some properties of packets, while the other roles such as error controls (detection and correction) and frame synchronization are omitted. The Physical Layer only transfers packets from a node to links, disregarding serialization and de-serialization of packets.

This reduction model has some entities [5] including source nodes (hosts), intermediate nodes (switches), destination nodes (hosts) and links. These source nodes, intermediate nodes, and links contain some atomic elements, including source queue, exit buffer, entrance buffer and unidirectional way which are defined as follows: **Source queue**: an unlimited size queue contains every packet as soon as it is generated by the source node. Each source node has only one source queue. **Exit buffer (EXB)**: a limited size queue contains several packets. A source node (or intermediate node) has one (or several) EXB(s) where the outgoing packets wait until they could leave to move to the next node. **Entrance buffer (ENB)**: a limited size queue contains several packets. An intermediate node has several ENBs where the incoming packets may wait for an empty slot of EXB. **Unidirectional way**: a link has two unidirectional ways to transfer packets from node A to node B and vice versa. At most one packet can go through a unidirectional way at a moment.

Figure 1 illustrates the atomic elements of (a) a source node that contains only one source queue and one EXB; and (b) an intermediate node (switch). In Fig. 1, a green slot indicates that a packet is located over there, a white slot is an empty one.

The structure of an intermediate node (switch) is characterized by the number of neighbor nodes and by the size of ENB as well as EXB. Let k >= 2 be the number of neighbor nodes which this intermediate node connects to, and k is also the number of links that attach to this node. Each link needs an ENB and an EXB, therefore, an intermediate node has k exit buffers and k entrance buffers.

3.2 Automata Specification

In this subsection, we describe the behavior of our network simulation from the perspective of DES. We utilize the most common modeling formalism Automata

[10] to capture event-triggered transitions between discrete states of a DES. There are five deterministic finite automata to represent discrete transitions of five atomic elements: (a) packet, (b) source queue, (c) entrance buffer, (d) exit buffer and (e) unidirectional way.

List of events: An event is an instantaneous occurrence that may change the state of an atomic element. It is characterized by the start time when it begins, and the end time when it finishes. In the process of transferring packets, these events following will occur:

Event type (a): source node generates a packet.

Event type (b): source node moves a packet from the source queue to its EXB. These events could be divided into two subgroups: event type (b1) and event type (b2). Events of type (b1) occur when the source queue has only one packet while events of type (b2) occur when source queue has more than one packet. Besides, events of type (b) could be categorized as type (b3) and (b4). The difference between two types is that if the ended event is of type (b3), the EXB still has at least one empty slot, otherwise, the EXB is full.

Event type (c): the source node removes the packet from the EXB to transfer.

Event type (d): link transfers the packet to ENB of the next intermediate node (switch). There are two subtypes of these events: (d1) and (d2). If an event of type (d1) is ended, the ENB is not full, otherwise, the ENB is full.

Event type (e): switch moves packet from ENB to EXB. This type could be categorized as two subtypes (e1) and (e2). If the ended event is of type (e1), the EXB still has at least an empty slot, otherwise, the EXB is full.

Event type (f): switch removes packet from the EXB to transfer.

Event type (g): link transfers packet to its destination.

Event type (h): a switch A receives notification from its neighbor B that from now on, ENB of B has a new empty slot.

State transition graph: Fig. 2 describes automata by using state transition graphs, in which nodes represent the states, and the arcs connecting the nodes represent all the possible transitions between states that are caused by the occurrence of events [10].

States of Packet:

- State P1: the packet is generated.
- State P2: the packet is located at EXB of the source node.
- State P3: the packet is moved in a unidirectional way.
- State P4: the packet is located at ENB of switch.
- State P5: the packet is located at EXB of switch.
- State P6: the packet is received by the destination node.

States of Source Queue:

- State Sq1: source queue is empty.
- State Sq2: source queue is not empty.

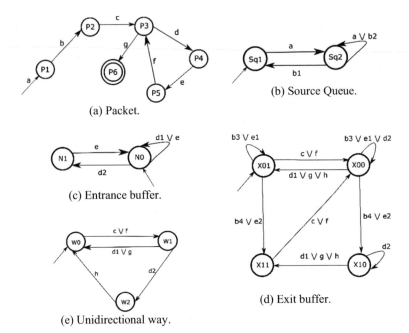

(a) Packet.

(b) Source Queue.

(c) Entrance buffer.

(d) Exit buffer.

(e) Unidirectional way.

Fig. 2 State-transition graphs of automata

States of ENB:

- State N0: ENB is not full.
- State N1: ENB is full.

States of EXB:

- State X01: EXB is not full and able to transfer packet.
- State X00: EXB is not full and unable to transfer packet.
- State X10: EXB is full and unable to transfer packet.
- State X11: EXB is full and able to transfer packet.

States of Unidirectional Way:

- State W0: the way has no packet and it is able to transfer one.
- State W1: the way has a packet.
- State W2: the way has no packet but it is unable to transfer.

Rule generation of events: Apart from events of type (h), an event is assigned with a packet, namely P. Event occurs if the atomic elements have properties and states which satisfy special conditions, in the following:

Event type (a) occurs if the source node generates packet P. The location where this event occurs is the source node; the start time is equal to the end time.

Event type (b) occurs if the source queue is at state Sq2, EXB is at state X01 or X00. The location of this event is the source node; the start time is equal to the end time.

Event type (c) occurs if the unidirectional way is at state W0, packet P is at the top of EXB, and EXB is at state X11 or X01. The location of this event is source node, the start time is equal to the end time.

Event type (d) occurs as soon as an event (c) or an event (f) ends. This event occurs in a unidirectional way, in which it starts when the packet begins to be transferred and ends when the packet reaches ENB of the next switch.

Event type (e) occurs if (i) packet P is on the top of ENB of the switch, (ii) EXB that the packet P needs to reach is at state X01 or X00, and (iii) the switch chooses packet P to move. This event occurs at that switch, it starts when the packet P is on top of ENB and ends after a SWITCH_CYCLE.

Event type (f) occurs if EXB is not empty, EXB is at state X01 or X11 and the unidirectional way is at state W0. This event occurs at the switch; it starts as the packet P is on top of EXB and ends after a SWITCH_CYCLE.

Event type (g) occurs as soon as an event (c) or an event (f) ends but the next node is destination one. This event occurs in a unidirectional way, in which it starts as the packet P is transferred from EXB of the neighbor switch to the destination node and ends as packet P reaches destination.

If node A has a connection to switch B; an event of type (h) occurs as ENB of switch B tranfers its top packet to an EXB of switch B, and the unidirectional way (from B to A) is at state W0. The start of this event along with switch B sends a notification to node A, and the event ends after one CYCLE.

4   Experimental Result

Our proposed model and specification has been implemented to produce a specialized simulator tool, using design approach in [12]. By using our specialized simulator, we reproduce the evaluation results of [8], in which the value of throughput in the case of the Fat-tree 4-port network is published. More specifically, we choose the communication pair which is a random pattern in combination with a Two-level Table. In Fig. 3, our experimental result shows that the throughput is 74%, almost the same to that as obtained in [8], 75%. The throughput performance of Fat-tree 4-port would be undermined if using NS3 for evaluation instead, therein, the maximum throughput is just 60%, significantly lower.

Last but not least, we illustrate simulation of a Fat Tree Network along with multiple runs as k ranges from 20 to 30. To evaluate the execution time of our specialized simulator and NS3, we use a computer that has Intel Core i7 2.3 GHz with 8 GB of main memory. Figure 4 depicts that our specialized simulator spends 2,564 s for simulating with k = 30 (in this case network has k*k*k/4 = 6,750 hosts), while NS3, however, requires about four hours.

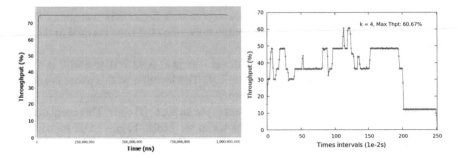

Fig. 3 Evaluation of our specialized tool (left) and NS3 (right)

Fig. 4 Execution time of
our specialized tool and NS3

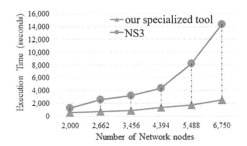

5   Conclusion

As they focus on simulating the full OSI 7-layer protocol stack, most popular network
simulation tools such as NS3 and Omnet++ are not fast enough for use in evaluating interconnection networks with a large size. To evaluate the performance of
network topologies, routing algorithms and congestion control strategies, researchers
have a tendency to make use of their own reduction network simulations. These
tools, however, often have no formal specification on their high-level abstraction.
We propose a reduction model to simulate packet transfer process in a large-scale
interconnection network, and deterministic finite automata to specify this model.
We implement our proposed reduction model and produce a specialized simulator,
which allows us to do the performance evaluation of large interconnection networks
much faster than using popular simulators such as NS3. Our experimental results
also reconfirm the self-evaluations of the landmark work [8], while the throughput
performance of this famous Fat-tree topology would have been undermined by using
the NS3 simulator instead (down to only 60 from 75% as in [8]). Our future work is to
implement parallelization for this framework to accelerate simulation in large-scale
interconnection networks.

References

1. Henderson TR, Roy S, Floyd S, Riley GF (2006) NS-3 project goals. In: Proceeding from the 2006 workshop on NS-2: the IP network simulator. ACM, New York, NY, USA, pp 13–20
2. Varga A, Hornig R (2008) An overview of the OMNeT++ simulation environment. In: Proceedings of the first international conference on simulation tools and techniques for communications, networks and systems, Article No. 60. ICST, Brussels, BEL, pp 1–10
3. Barr R, Haas ZJ, van Renesse R (2005) JiST: an efficient approach to simulation using virtual machines. Softw: Pract Experience, 35(6):539–576
4. Breslau L, Estrin D, Fall K, Floyd S, Heidemann J, Helmy A, Huang P, McCanne S, Varadhan K, Xu Y, Yu H (2000) Advances in network simulation, vol. 33, no. 5. IEEE, Computer, pp 59–67
5. Banks J, Carson II, Nelson BL, Nicol DM (2005) Discrete-event system simulation, 3rd edn. Pearson Education India, India
6. Weingartner E, Vom Lehn H, Wehrle K (2009) A performance comparison of recent network simulators. In: 2009 IEEE international conference on communications. IEEE, Dresden, Germany, pp 1–5
7. Brugge J, Paquereau L, Heegaard PE (2010) Experience report on implementing and simulating a routing protocol in ns-2 and ns-3. In: 2010 second international conference on advances in system simulation. IEEE, Nice, France, pp 88–93, 22 August 2010
8. Al-Fares M, Loukissas A, Vahdat A (2008) A scalable, commodity data center network architecture. In: ACM SIGCOMM computer communication review, vol 38, no 4. ACM, New York, NY, USA, pp 63–74, 17 August 2008
9. Al-Fares M, Radhakrishnan S, Raghavan B, Huang N, Vahdat A (2010) Hedera: dynamic flow scheduling for data center networks. In: Proceedings of the 7th USENIX conference on networked systems design and implementation, vol 10. USENIX Association, San Jose, California, USA, pp 89–92
10. Lafortune S (2019) Discrete event systems: modeling, observation, and control. Ann Rev Control, Robot, Auton Syst 2:141–159
11. Jyothi SA, Singla A, Godfrey PB, Kolla A (2016) Measuring and understanding throughput of network topologies. In: SC'16: Proceedings of the international conference for high performance computing, networking, storage and analysis. IEEE, Salt Lake City, UT, USA, pp 761–772
12. Chung KT, Thanh NT, Khanh-Van N (2019) An approach to designing software tool specified for evaluating performance of inter-connection topologies with large sizes. J Res Develop Inf Commun Technol 2019(1), Vietnamese Edition, pp 27–38, ICT Research, Vietnam

Interworking of 5G and Broadcasting Network Using 5G Network Slicing

Hyuncheol Kim, Hye Ju Oh, and Soon Choul Kim

Abstract Over the last few years, terrestrial and pay-TV environments have evolved into a more competitive broadcasting service infrastructure environment with the IP paradigm shift and the introduction of cloud virtualization. To cope with such changes in the broadcasting industry, the dependence of cloud virtualization on the broadcasting infrastructure will increase, and high value-added broadcasting services will continue to emerge in the broadcasting environment that can guarantee efficient resource use and management. To have these features, many challenges must be solved. In this paper, we introduce the scheme and the recent research trend for interworking 5G network and broadcasting network using a 5G network slicing function.

Keywords 5G · Broadcasting network · 5G interworking

1 Introduction

Over the last few years, terrestrial and pay-TV environments have evolved into a more competitive broadcasting service infrastructure environment with the IP paradigm shift and the introduction of cloud virtualization. To cope with such changes in the broadcasting industry, the dependence of cloud virtualization on the broadcasting infrastructure will increase, and high value-added broadcasting services will continue to emerge in the broadcasting environment that can guarantee efficient resource use and management.

H. J. Oh (✉) · S. C. Kim
Electronics and Telecommunications Research Institute, Daejeon, South Korea
e-mail: feeler@etri.re.kr

S. C. Kim
e-mail: choulsim@etri.re.kr

H. Kim
Namseoul University, Cheonan, South Korea
e-mail: hckim@nsu.ac.kr

© The Author(s), under exclusive license to Springer Nature Singapore Pte Ltd. 2021
H. Kim et al. (eds.), *IT Convergence and Security*, Lecture Notes
in Electrical Engineering 712, https://doi.org/10.1007/978-981-15-9354-3_19

To have these features, many challenges must be solved. Not only traditional technical issues such as mobility support and session management, which have been considered important issues in mobile communication network design until now. Separation of network functions, separation of user and control planes to provide a highly flexible environment, separation of boundaries between network domains (RAN and core networks, transport, and mobile cores, networks and services, etc.), orchestration of virtualized network functions. There are several new issues associated with life cycle management (LCM), allocation of infrastructure resources, and so on [1, 2].

Research on designing 5G network architectures to accommodate the new paradigm, including added complexity (many new interfaces, new network management functions, security issues), is still in the process of abstracting the concepts. Each researcher proposes a 5G network system structure from a different point of view based on his technical expertise. There is a big difference in the system structure that the traditional network equipment industry, broadcasting equipment industry, and computing software industry including network function virtualization (NFV) and software-defined network (SDN). Since 5G networks must be a convergence of network and computing, a rational structure must be derived, not a contest of initiatives between the two camps [3–5].

In this paper, we introduce the scheme and the recent research trend for interworking 5G network and broadcasting network using a 5G network slicing function.

2   (4)5G and (F)eMBMS

2.1   4G LTE and eMBMS

Since 3G mobile communication technology was developed, the rapid increase in smartphone penetration has led to the spread and demand for various types of data services such as video streaming beyond photo and text transmission. In particular, as the consumption of video content in the mobile environment continues to increase, multimedia services including video have become the main service area of mobile communication systems. Also, video content consumption continued to increase due to network quality improvements, handset size and resolution improvements, and high definition content (HD). However, video content transmission through a mobile network may be limited to use the service if data use is concentrated in a certain region. Indeed, in 2011, NHN provided a real-time baseball game using 3G networks, but traffic surged as more than 20,000 viewers gathered at once. NHN changed the service to Wi-Fi only relaying within one month of the implementation of the service. The biggest reason for the problem of video content transmission through the mobile network is because mobile communication technology is based on unicast, and the load on the network increases as the number of users increases. Accordingly,

Fig. 1 MBMS network architecture

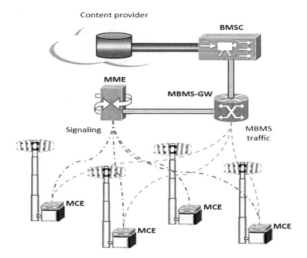

eMBMS (evolved Multimedia Broadcast & Multicast Service) technology has been proposed to overcome the limitations of video content transmission in LTE networks and to satisfy various data service needs.

eMBMS, as shown in Fig. 1, is based on broadcast, a one-to-many transmission method. It is also called LTE Broadcast, a technology that can transmit the same data to an unspecified number of users through an LTE network. Unlike the one-to-one unicast transmission method of the existing mobile communication, the broadcast is transmitted in a one-to-many manner to provide a service to a large number of users without a network load, and the efficiency of the network is also increased. That is, the eMBMS technology can provide a service without any load on the entire network even if the number of users increases because many users receive the same data at once. A key element of the eMBMS is a multicast-broadcast single frequency network (MBSFN) transmission technology capable of simultaneously transmitting the same data (content) from different base stations. With MBSFN technology, multiple cells can operate as one large cell by simultaneously transmitting the same data across multiple adjacent cells.

3 5G and Broadcasting Network Interworking

3.1 Network Slicing

'Network Slicing' is a technology that separates one physical network infrastructure into multiple independent virtual networks according to service types and provides a dedicated network specialized for each service for various services having different characteristics. Each network slice is guaranteed virtualized network resources and

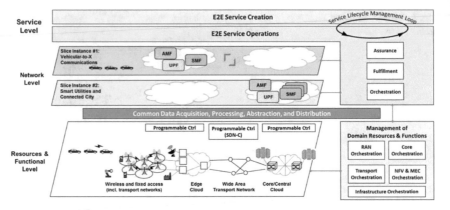

Fig. 2 5G Enhanced Overall System Architecture (5G PPP)

is disconnected from each other so that an error or failure in one slice does not affect the communication of the other slices.

Slice means a piece. By dividing the network into pieces, you can prevent problems with one service from failing to another. Of course, if traffic increases in one piece, you can use other pieces of resources. A network slice refers to a network in which base stations and exchanges are logically integrated. For example, you might have a service that needs to be very fast, and there may be a service that doesn't have to be as fast as the IoT, but that requires multiple simultaneous connections. Network slicing is a concept created to more efficiently use network resources to meet the requirements of these services [6, 7].

3.2 Network Slice Generation and Interworking

Figure 2 shows the end to end (ETE) network slicing service provision structure proposed by 5G PPP and the general 5G network. Therefore, the ATSC 3.0 headend requires a new module that can work with 5G to create 5G network slices, as shown in Fig. 3.

4 Conclusion

The broadcasting network has grown in a separate market based on technology separate from telecommunication fields. Following the trend of digitalization of global broadcasting, major countries have made their efforts to introduce digital broadcasting and revise the system to activate related industries. Through this process, each country's broadcasting market also has unique structural characteristics.

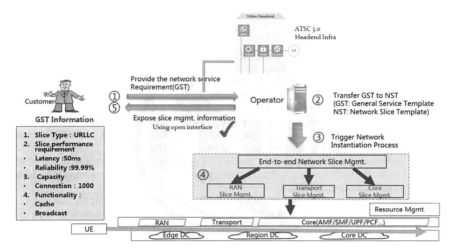

Fig. 3 ATSC 3.0 head-end architecture for 5G integration

During such changes, TV transmission technologies, which have been evolving continuously, have recently developed to a level that can revolutionize the market. In other words, the terrestrial broadcasting standard ATSC 3.0, which was developed in 2015, was designed with the convergence of 5G, a mobile communication technology in mind, and both technologies are based on Internet protocol. As a result, it is expected that the transmission method may be selectively used among broadcast, multicast, and unicast according to the number of broadcast contents users. This means that it is possible to provide a video service by selecting an efficient technology that is low in cost depending on the number of users.

Therefore, under 5G and IP-based broadcasting networks, it would be desirable to prevent obstacles in properly using cost-effective technologies according to the number and method of broadcasting content users. This will be possible by harmonizing the network of broadcasting and telecommunications, and for this, discussion and plans for sound development and propagation promotion of the entire ecosystem of broadcasting and telecommunications will be necessary.

Acknowledgments This work was supported by Institute of Information & communications Technology Planning & Evaluation (IITP) grant funded by the Korea government (MSIT) (No. 2018-0-00786, Development of cloud virtualization technology for broadcasting system).

References

1. Gimenez JJ, Gomez-Barquero D, Morgade J, Stare E (2018) Wideband broadcasting: a power-efficient approach to 5G broadcasting. IEEE Commun Mag 3(56):119–125
2. Gimenez JJ, Carcel JL, Fuentes M, Garro E, Elliott S, Vargas D, Menzel C, Gomez-Barquero D (2019) Wideband broadcasting: 5G new radio for terrestrial broadcast: a forward-looking approach for NR-MBMS. IEEE Trans Broadcast 2(65):356–368
3. Heyn T, Morgade J, Petersen S, Pfaffinger K, Lang E, Hertlein M, Fischer G (2016) Integration of broadcast and broadband in LTE/SG (IMBS)-experimental results from the eMBMS Testbeds. In: European conference on networks and communications (EuCNC)
4. Guo W, Fuentes M, Christodoulou L, Mouhouche B (2018) Roads to multimedia broadcast multicast services in 5G new radio. In: IEEE international symposium on broadband multimedia systems and broadcasting (BMSB)
5. Kourtis MA, Blanco B, Pérez-Romero J, Makris D, McGrath MJ, Xilouris G, Munaretto D, Solozabal R, Sanchoyerto A, Giannoulakis I, Kafetzakis E (2019) A cloud-enabled small cell architecture in 5G networks for broadcast/multicast services. IEEE Trans Broadcast 2(65):414–424
6. Wang Y, He D, Ding L, Zhang W, Li W, Yiyan Wu, Liu N, Wang Y (2017) Media transmission by cooperation of cellular network and broadcasting network. IEEE Trans Broadcast 3(65):571–576
7. Öztürk E, Zia W, Pauli V, Steinbach E (2018) Performance evaluation of ATSC 3.0 DASH Over LTE eMBMS. In: IEEE international symposium on broadband multimedia systems and broadcasting (BMSB)

Printed in the United States
by Baker & Taylor Publisher Services